Harcourt Math

Practice for the ISAT Test

TEACHER EDITION
Grade 4

Harcourt

Orlando   Austin   New York   San Diego   Toronto   London

Visit *The Learning Site!*
www.harcourtschool.com

Copyright © by Harcourt, Inc.

All rights reserved. No part of this publication may be reproduced or transmitted in any form or by any means, electronic or mechanical, including photocopy, recording, or any information storage and retrieval system, without permission in writing from the publisher.

Permission is hereby granted to individual teachers using the corresponding student's textbook or kit as the major vehicle for regular classroom instruction to photocopy Copying Masters from this publication in classroom quantities for instructional use and not for resale. Requests for information on other matters regarding duplication of this work should be addressed to School Permissions and Copyrights, Harcourt, Inc., 6277 Sea Harbor Drive, Orlando, Florida 32887-6777. Fax: 407-345-2418.

HARCOURT and the Harcourt Logo are trademarks of Harcourt, Inc., registered in the United States of America and/or other jurisdictions.

Printed in the United States of America

ISBN 0-15-341783-8

1 2 3 4 5 6 7 8 9 10 082 13 12 11 10 09 08 07 06 05 04

CONTENTS

Overview iv

Harcourt Math—a K–6 Math Program
Published by Harcourt School Publishers v

Fast Track Pacing Guide vi

Alternative Unit Sequences x

Communicating Ideas in Mathematics xi

Good Questions to Ask xii

Problem-Solving Think Along xiii

Tips for Answering Multiple-Choice Items xiv

Scoring Extended-Response Tasks with a Rubric xv

Mathematics Scoring Rubric xviii

Individual Record Forms xix

Vocabulary Practice 1

Pretest 6

Units 1–9 Practice Tests 11

Posttest 38

Standards Correlation 43

Answer Sheet 48

OVERVIEW

Practice for the ISAT Test

Fast Track Pacing Guide
In the Teacher's Edition, a chart is provided that lists essential *Harcourt Math* lessons to teach before students take the ISAT.

Unit Tests
Nine unit tests are provided for Grades 3–5, and six unit tests are provided for Grade 2. Each unit test includes multiple-choice and extended-response items that practice Illinois Learning Standards covered in chapters of *Harcourt Math*. Frequent practice of applying prior learning in a problem-solving setting will enable all students to show what they know when they take the test.

Practice Tests
Two practice tests are provided. Each test includes items that practice content from the Illinois Learning Standards for Mathematics. The first practice test is a Pretest. The second practice test, or Posttest, can be given as testing practice before the ISAT is administered. At each grade level, items are in the same format as they will appear on the Mathematics ISAT.

Individual Record Form
The Individual Record Form allows you to record and track students' progress on the Pretest and Posttest. The form is organized by strands to indicate which objectives are tested. The form also correlates each item to a lesson in *Harcourt Math* in which the objective is taught.

Vocabulary Practice
Vocabulary practice for each of the five strands is provided. These engaging activities provide opportunities to reinforce students' understanding of mathematical concepts.

Harcourt Math—a K–6 Math Program
Published by Harcourt School Publishers

Harcourt Math provides balanced instruction in concepts, skills, and problem-solving strategies and applications to ensure a solid foundation in every strand as defined in the Illinois Learning Standards for Mathematics.

How does *Harcourt Math* prepare your students for the ISAT?

***Harcourt Math* fully covers the curriculum outlined in the Illinois Learning Standards for Mathematics.**

Your students will answer ISAT questions confidently and correctly because they have learned and practiced

- ▶ concepts and skills.
- ▶ problem-solving strategies and applications.
- ▶ mathematical reasoning.
- ▶ communication about mathematics.

***Harcourt Math* effectively teaches students how to respond in writing to test questions.**

Students will know how to communicate their reasoning because they have used

- ▶ Talk About It and Write About It.
- ▶ Performance assessment tasks in every unit with self-assessment tools.
- ▶ Problem-Solving Think Along.

***Harcourt Math* Assessment Components support teaching and learning in all formats.**

- ▶ Performance Assessment (1–6)
- ▶ Assessment Guide: test copying masters (in free response and multiple-choice formats)
- ▶ Harcourt Assessment System CD-ROMs (1–6)
- ▶ *Practice for the ISAT Test* (student edition and teacher edition)

Introduction   **Practice for the ISAT Test**

Fast Track Pacing Guide

The Fast Track Pacing Guide shows critical content to teach before the Mathematics ISAT, making it easy to teach standards-based lessons for ISAT success.

		Grade 4 *Harcourt Math* Lesson	Standards
Week 1	\multicolumn{2}{l	}{PRETEST, pp. 11 – 20}	
	1.1	Understand Place Value	6.4.03 (A)
	1.2	Place Value Through Hundred Thousands	6.4.03 (A); 6.4.04 (A); 6.4.22 (A); 6.4.24 (A)
	1.4	Benchmark Numbers	6.4.18 (A)
	2.1	Compare Numbers	6.4.02 (A); 6.4.03 (A); 6.4.04 (A); 6.4.25 (A)
Week 2	2.2	Order Numbers	6.4.02 (A); 6.4.03 (A); 6.4.04 (A); 6.4.23 (A)
	2.4	Round Numbers	6.4.17 (B/C)
	3.2	Estimate Sums and Differences	6.4.15 (B/C); 6.4.20 (A)
	3.3	Add and Subtract to 4-Digit Numbers	6.4.16 (B/C)
	3.4	Subtract Across Zeros	6.4.16 (B/C)
Week 3	3.6	Problem Solving Skill: Estimate or Find Exact Answers	6.4.15 (B/C); 6.4.19 (B/C)
	4.1	Expressions	8.4.03 (C/D)
	4.3	Equations	6.4.28 (B/C); 8.4.03 (C/D)
	4.4	Patterns: Find a Rule	8.4.01 (A)
	4.5	Hands On: Balance Equations	6.4.28 (B/C); 8.4.07 (A)
Week 4	\multicolumn{2}{l	}{Unit 1 Test}	
	5.2	Elapsed Time	7.4.02 (A/C)
	6.1	Collect and Organize Data	10.4.01 (A/B)
	6.2	Hands On: Find Mean, Median, and Mode	10.4.04 (A/B)
	6.4	Make Stem-and-Leaf Plots	10.4.02 (A/B)
Week 5	6.5	Compare Graphs	10.4.03 (A/B)
	6.6	Problem Solving Strategy: Make a Graph	10.4.02 (A/B)
	7.1	Hands On: Make Bar and Double-Bar Graphs	8.4.02 (A)
	7.2	Read Line Graphs	8.4.02 (A); 10.4.01 (A/B)
	7.3	Hands On: Make Line Graphs	8.4.02 (A); 10.4.02 (A/B)

Suggested Pacing: 1 day for each row in the table

Practice for the ISAT Test — **Introduction**

		Grade 4 *Harcourt Math* Lesson	**Standards**
Week 6	7.4	Hands On: Make Circle Graphs	10.4.02 (A/B)
	7.5	Choose an Appropriate Graph	8.4.02 (A)
	7.6	Problem Solving Skill: Draw Conclusions	10.4.01 (A/B); 10.4.06 (A/B)
	Unit 2 Test		
	8.1	Relate Multiplication and Division	6.4.21 (B/C)
Week 7	8.2	Multiply and Divide Facts Through 5	6.4.16 (B/C); 6.4.26 (B/C)
	8.3	Hands On: Multiply and Divide Facts Through 10	6.4.16 (B/C); 6.4.21 (B/C); 6.4.26 (B/C)
	8.4	Hands On: Multiplication Table Through 12	6.4.16 (B/C); 6.4.26 (B/C)
	9.2	Hands On: Order of Operations	8.4.05 (C/D); 8.4.06 (C/D)
	9.3	Expressions and Equations with Variables	8.4.03 (C/D); 8.4.06 (C/D)
Week 8	9.5	Hands On: Balance Equations	8.4.06 (C/D); 8.4.07 (A)
	9.6	Patterns: Find a Rule	8.4.01 (A); 9.4.10 (A)
	Unit 3 Test		
	10.2	Estimate Products	6.4.15 (B/C)
	10.3	Hands On: Multiply 2-Digit Numbers	6.4.16 (B/C)
Week 9	10.4	Multiply 3- and 4-Digit Numbers	6.4.16 (B/C)
	11.3	Multiply by Tens	6.4.16 (B/C); 6.4.26 (B/C)
	11.4	Estimate Products	6.4.15 (B/C)
	11.5	Problem Solving Strategy: Solve a Simpler Problem	8.4.04 (A)
	12.1	Multiply 2-Digit Numbers	6.4.16 (B/C)
Week 10	12.2	Multiply 3-Digit Numbers	6.4.16 (B/C)
	12.5	Problem Solving Skill: Multistep Problems	8.4.04 (A)
	Unit 4 Test		
	13.1	Hands On: Divide with Remainders	6.4.16 (B/C)
	14.1	Estimate Quotients	6.4.15 (B/C)
Week 11	14.3	Divide 3-Digit Numbers	6.4.16 (B/C)
	14.4	Zeros in Division	6.4.16 (B/C)
	15.3	Division Procedures	6.4.16 (B/C)
	15.5	Problem Solving Skill: Choose the Operation	6.4.21 (B/C)
	16.2	Hands On: Factors and Multiples	6.4.09 (A)

Suggested Pacing: 1 day for each row in the table

Introduction

		Grade 4 Harcourt Math Lesson	**Standards**
Week 12	16.3	Hands On: Prime and Composite Numbers	6.4.07 (A); 6.4.08 (A)
	16.5	Square Numbers	6.4.09 (A)
	Unit 5 Test		
	18.1	Hands On: Polygons	9.4.02 (A); 9.4.04 (B); 9.4.09 (A)
	18.2	Classify Triangles	9.4.09 (A)
Week 13	18.3	Hands On: Classify Quadrilaterals	9.4.09 (A)
	18.5	Hands On: Circles	9.4.03 (A)
	20.1	Temperature: Fahrenheit	7.4.04 (A/C); 7.4.10 (A/C)
	20.2	Temperature: Celsius	7.4.04 (A/C); 7.4.10 (A/C)
	20.6	Use a Coordinate Grid	9.4.07 (A); 9.4.08 (A)
Week 14	Unit 6 Test		
	21.1	Read and Write Fractions	6.4.05 (A); 6.4.27 (A)
	21.2	Hands On: Equivalent Fractions	6.4.11 (D); 6.4.12 (D)
	21.3	Hands On: Compare and Order Fractions	6.4.01 (A); 6.4.11 (D)
	21.4	Problem Solving Strategy: Make a Model	6.4.11 (D); 6.4.27 (A)
Week 15	21.5	Mixed Numbers	6.4.05 (A); 6.4.11 (D)
	22.1	Add Like Fractions	6.4.13 (B/C); 6.4.18 (A)
	22.2	Hands On: Subtract Like Fractions	6.4.13 (B/C)
	23.2	Predict Outcomes of Experiments	10.4.05 (C)
	Unit 7 Test		
Week 16	24.1	Hands On: Length: Choose the Appropriate Unit	7.4.06 (B)
	24.2	Hands On: Measure Fractional Parts	7.4.01 (A/C) (OR)
	24.3	Algebra: Change Linear Units	7.4.03 (A/C) (OR); 7.4.09 (A/C)
	24.4	Hands On: Capacity	7.4.04 (A/C); 7.4.08 (B)
	24.5	Hands On: Weight	7.4.04 (A/C); 7.4.08 (B)

Suggested Pacing: 1 day for each row in the table

Practice for the ISAT Test   **Introduction**

		Grade 4 *Harcourt Math* Lesson	**Standards**
Week 17	24.6	Problem Solving Strategy: Compare Strategies	7.4.07 (B)
	25.1	Hands On: Metric Length	7.4.04 (A/C); 7.4.08 (B)
	25.2	Algebra: Change Linear Units	7.4.03 (A/C) (OR); 7.4.09 (A/C)
	25.3	Hands On: Capacity	7.4.04 (A/C); 7.4.08 (B)
	25.4	Hands On: Mass	7.4.04 (A/C); 7.4.08 (B)
Week 18	25.6	Relate Benchmark Measurements	7.4.03 (A/C) (OR)
	26.1	Relate Fractions and Decimals	6.4.05 (A); 6.4.11 (D)
	26.2	Decimals to Thousandths	6.4.05 (A); 6.4.06 (A); 6.4.10 (A)
	26.3	Hands On: Equivalent Decimals	6.4.10 (A); 6.4.11 (D); 6.4.12 (D)
	26.4	Relate Mixed Numbers and Decimals	6.4.11 (D)
Week 19	26.5	Compare and Order Decimals	6.4.02 (A)
	27.2	Estimate Decimal Sums and Differences	6.4.15 (B/C)
	27.3	Hands On: Add Decimals	6.4.14 (B/C); 6.4.29 (B/C)
	27.4	Hands On: Subtract Decimals	6.4.14 (B/C); 6.4.29 (B/C)
	27.5	Add and Subtract Decimals and Money	6.4.14 (B/C); 6.4.18 (A); 6.4.29 (B/C)
Week 20	Unit 8 Test		
	28.2	Hands On: Estimate and Find Perimeter	7.4.05 (A/C) (OR)
	29.2	Find Area	7.4.05 (A/C) (OR)
	29.3	Hands On: Relate Area and Perimeter	7.4.04 (A/C)
	29.4	Problem Solving Strategy: Find a Pattern	7.4.04 (A/C)
Week 21	30.1	Hands On: Faces, Edges, and Vertices	9.4.01 (A); 9.4.06 (B)
	30.2	Hands On: Patterns for Solid Figures	9.4.05 (B)
	Unit 9 Test		
	POSTTEST, pp. 75 – 84		

Suggested Pacing: 1 day for each row in the table

Introduction — **Practice for the ISAT Test**

Alternative Unit Sequences
Harcourt Math—Grade 4

Harcourt Math presents topics in a sequence that mathematicians and authors who guided the development of the program considered to be the most mathematically appropriate.

There are certain topics, however, that offer flexibility for alternate sequencing. That is, they can be taught in almost any order. These include some topics in geometry, measurement, and data.

Listed below are the present sequence of units, as reflected in the Table of Contents of *Harcourt Math*, along with two alternative sequences you may wish to use.

PRESENT SEQUENCE OF UNITS	ALTERNATIVE SEQUENCE A	ALTERNATIVE SEQUENCE B
Unit 1 Understand Whole Numbers and Operations	**Unit 1** Understand Whole Numbers and Operations	**Unit 2** Time, Data, and Graphing
Unit 2 Time, Data, and Graphing	**Unit 3** Multiplication and Division Facts	**Unit 1** Understand Whole Numbers and Operations
Unit 3 Multiplication and Division Facts	**Unit 4** Multiply by 1- and 2-Digit Numbers	**Unit 3** Multiplication and Division Facts
Unit 4 Multiply by 1- and 2-Digit Numbers	**Unit 5** Divide by 1- and 2-Digit Divisors	**Unit 4** Multiply by 1- and 2-Digit Numbers
Unit 5 Divide by 1- and 2-Digit Divisors	**Unit 2** Time, Data, and Graphing	**Unit 5** Divide by 1- and 2-Digit Divisors
Unit 6 Geometry and Algebra	**Unit 7** Fractions and Probability	**Unit 6** Geometry and Algebra
Unit 7 Fractions and Probability	**Unit 8** Measurement and Decimals	**Unit 7** Fractions and Probability
Unit 8 Measurement and Decimals	**Unit 9** Perimeter, Area, and Volume	**Unit 8** Measurement and Decimals
Unit 9 Perimeter, Area, and Volume	**Unit 6** Geometry and Algebra	**Unit 9** Perimeter, Area, and Volume

Practice for the ISAT Test   **Introduction**

Communicating Ideas in Mathematics

The ISAT items in mathematics ask students to explain their thinking by:
- ▶ describing a problem-solving procedure or
- ▶ interpreting a data set or the results of a survey or
- ▶ justifying a method used to solve a problem.

Reading, summarizing, and visualizing are all key communication skills. You already use them in language arts. Make them work for you and your students in mathematics instruction as well.

Communicating About Mathematics

Students learn to:
- ▶ identify the question they are to answer.
- ▶ show what they know.
- ▶ clarify their understanding.
- ▶ check their solutions.
- ▶ develop flexible problem-solving skills.

Teachers learn:
- ▶ how students approach new ideas.
- ▶ how students connect prior learning to new concepts.
- ▶ how to use communication as an assessment tool.

Tips for Talking and Writing About Math

Help students think.
- ▶ Have students restate the problem in their own words.

Help students solve.
- ▶ Help students be good listeners. There is usually more than one good way to solve a problem.
- ▶ Students can learn from their mistakes. Have them describe their solution strategies and identify sources of error.

Help students explain.
- ▶ Have all students justify all answers, right or wrong. Encourage them to clearly explain their thinking, using numbers, symbols, words, and pictures.

Introduction — Practice for the ISAT Test

Good Questions to Ask

Use the Problem-Solving Think Along to help students develop an orderly approach to solving problems.

The Problem-Solving Think Along is available as a blackline master at p. xiii of the Teacher Edition to help guide class discussion.

Use it every time students work a word problem until they have internalized the steps to successful problem solving.

Problem-Solving Think Along

Understand
1. Retell the problem in your own words. _____
   _____
   _____
   _____

2. List the information given. _____
   _____

3. Restate the question as a fill-in-the-blank sentence. _____
   _____
   _____

Plan
4. List one or more problem-solving strategies that you can use.
   _____

5. Predict what your answer will be. _____
   _____

Solve
6. Show how you solved the problem. _____
   _____
   _____

7. Write your answer in a complete sentence. _____
   _____

Check
8. Tell how you know your answer is reasonable. _____
   _____

9. Describe another way you could have solved the problem.
   _____

It takes practice to become good communicators. Give your students daily opportunities to share ideas and discuss exemplary responses. Encourage students to use words and pictures, numbers and charts, to enhance their explanations and expand their problem-solving skills.

Name_____

> PROBLEM-SOLVING THINK ALONG

Problem-Solving Think Along

Understand
1. Retell the problem in your own words. _____
_____
_____
_____

2. List the information given. _____
_____

3. Restate the question as a fill-in-the-blank sentence. _____
_____
_____

Plan
4. List one or more problem-solving strategies that you can use.
_____

5. Predict what your answer will be. _____
_____

Solve
6. Show how you solved the problem. _____
_____
_____

7. Write your answer in a complete sentence. _____
_____

Check
8. Tell how you know your answer is reasonable. _____
_____

9. Describe another way you could have solved the problem.
_____

Introduction xiii **Practice for the ISAT Test**

Tips for Answering Multiple-Choice Items

Multiple-Choice Items

Multiple-choice test items will be familiar to students from other testing experiences. Five answer choices are given in the Grade 4 and Grade 5 ISATs. Four answer choices are given in the Grade 2 and Grade 3 ISATs. A test item itself may be a multistep problem.

To help students succeed with multiple-choice test items:

- ▶ Have students solve the problem first.
- ▶ Have them find the answer choice that matches their solution.
- ▶ After the correct answer has been identified, demonstrate how to fill in the answer bubble cleanly.

What if no answer matches?

- ▶ Have students restate the problem in their own words to determine the source of the error.
- ▶ Then have them evaluate each answer choice to eliminate ones that do not fit the problem.
- ▶ Have students work additional multiple-choice practice problems independently.
- ▶ Have volunteers demonstrate how they solved multiple-choice problems.

Response time: Students should try to complete a multiple-choice item in about 1 minute.

Scoring Extended-Response Tasks with a Rubric

Help Students Understand What Scorers Expect

1. Discuss the rubric with students.
2. Have students score their own answer to a practice task, using the rubric.
3. Discuss results. Have students revise their work to improve their score.

Help Students Understand How to Show What They Know

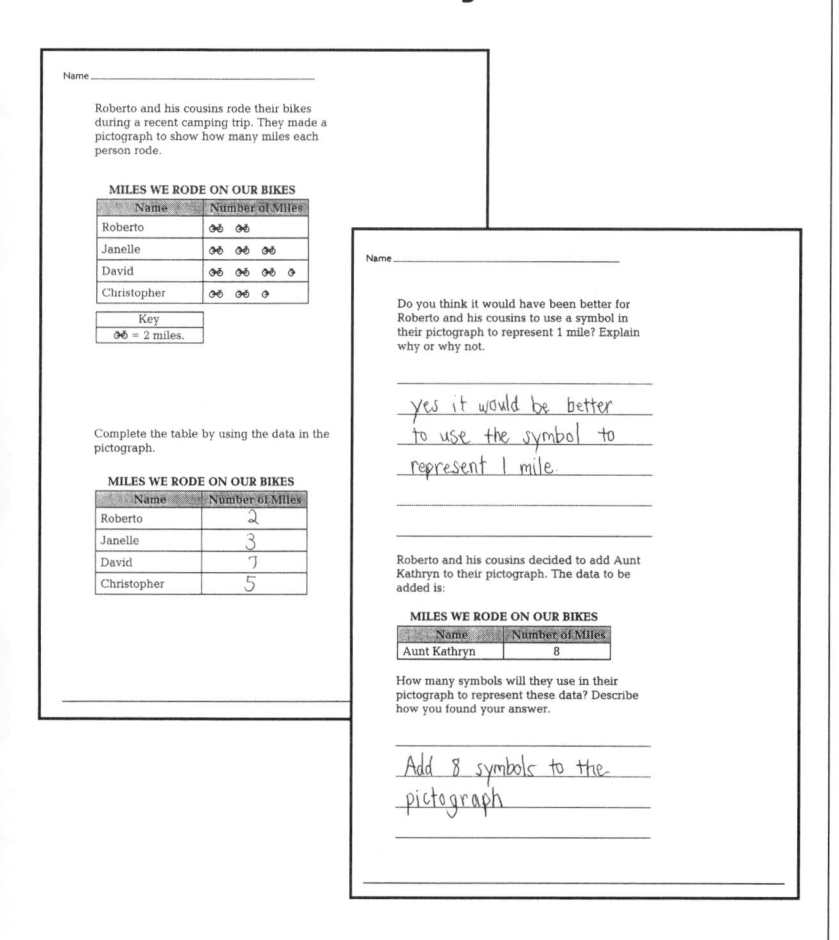

Have students evaluate this response to understand how it can be improved.

Ask, "Did the student answer each question?"

Help students see that the answers to the first part of the problem are complete but not all correct. On the second part of the problem, he did not explain his answer.

Partially correct response

Daily practice will help students *read each problem more carefully.*

Discussing solution strategies will help them learn to sort through the math knowledge needed to answer the required questions clearly and completely.

Introduction xv **Practice for the ISAT Test**

Scoring Extended-Response Tasks with a Rubric

Ask, "What can students do to improve their scores?"

A student can restate each part of the problem in their own words. Then students can check their restatements against the problem given and against their answers.

Have your students evaluate this exemplary response to understand why it is complete.

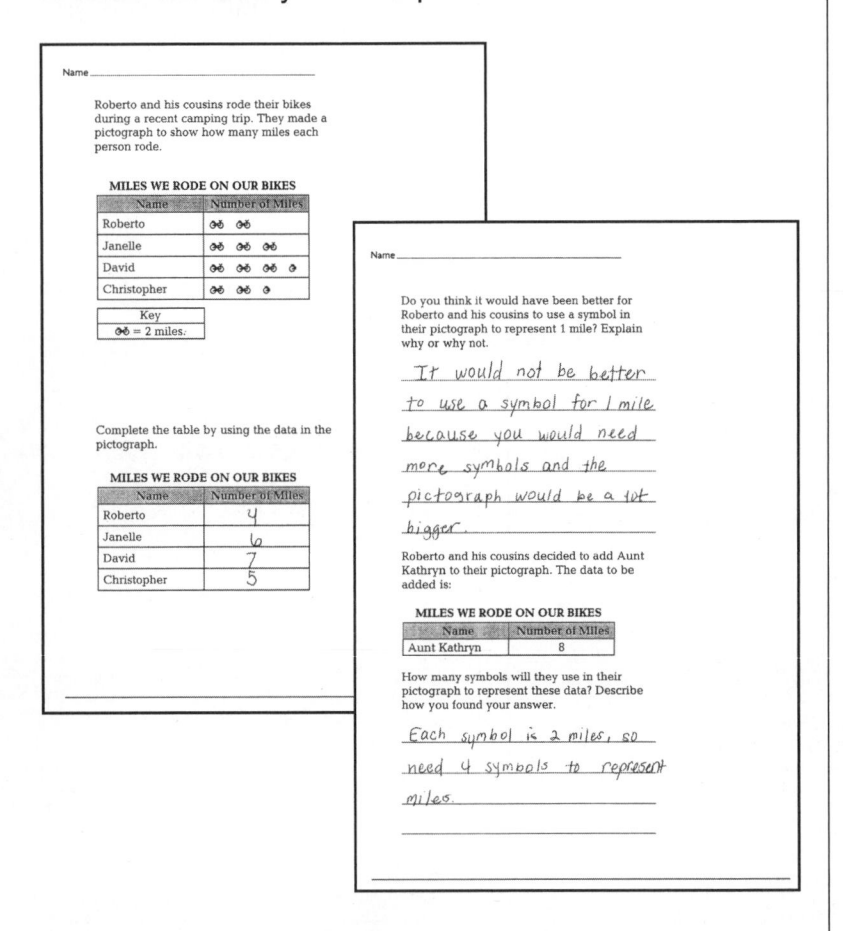

Exemplary response

This student has carefully read the problem. She has filled in the table and answered the questions correctly. The student has complete explanations.

Help Students Practice Reviewing and Revising Their Own Work

1. Have a volunteer share a response to a performance task question.
2. Have students discuss the answer.
3. Have students revise their own work to improve their scores.

Practice for the ISAT Test xvi **Introduction**

Scoring Extended-Response Tasks with a Rubric

Extended-response items are scored using a rubric. Students can receive partial credit for a partially completed or partially correct answer. Answers can earn from 0 to 4 points. Students can use 3–5 minutes to respond to extended-response test items.

Scoring Rubric

4 points   A score of 4 indicates that the student demonstrates a thorough understanding of key concepts and procedures. The student responds correctly to the task and provides clear and complete explanations and interpretations.

The response may contain minor flaws that do not detract from the demonstration of a thorough understanding.

3 points   A score of 3 indicates that the student demonstrates an understanding of key concepts and procedures. The student's response is essentially correct, with the mathematical procedures used and the explanations and interpretations provided demonstrating an essential but less than thorough understanding.

The response contains minor flaws in reasoning and/or computation or fails to address an aspect of the task.

2 points   A score of 2 indicates that the student demonstrates only a partial understanding of key concepts and procedures. Although the student may use the correct approach to obtain a solution or may provide a correct solution, the student's work lacks an essential understanding of the underlying mathematical concepts.

The response contains errors related to misunderstanding of relevant mathematical procedures or concepts or faulty interpretations of results.

1 point   A score of 1 indicates that the student demonstrates a very limited understanding of key concepts and procedures. The response is incomplete and exhibits many flaws.

0 points   A score of 0 indicates that the student has provided a completely incorrect or irrelevant response or no response at all.

Mathematics Scoring Rubric:
A Guide to Scoring Extended-Response Tasks

Score Level	MATHEMATICAL KNOWLEDGE: Knowledge of mathematical principles and concepts that result in a correct solution to a problem	STRATEGIC KNOWLEDGE: Identification of important elements of the problem and the use of models, diagrams, symbols, and/or algorithms, to systematically represent and integrate concepts	EXPLANATION: Written explanation and rationales that translate into words the steps of the solution process and provide justification for each step; though important, the length of the response, grammar, and syntax are not the critical elements of this dimension.
4	• shows complete understanding of the problem's mathematical concepts and principles • uses appropriate mathematical terminology and notations, including labeling answer if appropriate; that is, whether the unit is called for in the stem of the item • executes algorithms completely and correctly	• identifies all the important elements of the problem and shows complete understanding of the real relationships among elements • reflects an appropriate and systematic strategy for solving the problem • gives clear evidence of a complete and systematic solution process	• gives a complete written explanation of the solution process employed; explanation addresses both what was done and why it was done • if a diagram is appropriate, includes a complete explanation of all the elements in the diagram
3	• shows nearly complete understanding of the problem's mathematical concepts and principles • uses nearly correct mathematical terminology and notations • executes algorithms completely; computations are generally correct but may contain minor errors	• identifies most of the important elements of the problem and shows general understanding of the relationships among them • reflects an appropriate strategy for solving the problem • nearly complete solution process	• gives a nearly complete written explanation of the solution process employed; clearly explains what was done and begins to address why it was done • may include a diagram with most of the elements explained
2	• shows some understanding of the problem's mathematical concepts and principles • may contain major computational errors	• identifies some important elements of the problem but shows only limited understanding of the relationships among them • appears to reflect an appropriate strategy but the application of strategy is unclear, or a related strategy is applied logically and consistently • gives some evidence of a solution process	• gives some written explanation of the solution process employed; either explains what was done or addresses why it was done; explanation is vague or difficult to interpret • may include a diagram with some of the elements explained
1	• shows limited to no understanding of the problem's mathematical concepts and principles • may misuse or fail to use mathematical terms • may contain major computational errors	• fails to identify important elements or places too much emphasis on unimportant elements • may reflect an inappropriate or inconsistent strategy for solving the problem • gives minimal evidence of a solution process; process may be difficult to identify • may attempt to use irrelevant outside information	• gives minimal written explanation of solution process; may fail to explain what was done and why it was done • explanation does not match presented solution process • may include minimal discussion of elements in diagram; explanation of significant element is unclear
0	• no answer attempted	• no apparent strategy	• no written explanation of the solution process provided

Practice for the ISAT Test   xviii   Introduction

Individual Record Form

Name _____ Date _____

▶ PRETEST & POSTTEST

OBJECTIVES & LESSONS ASSESSED			PRETEST				POSTTEST				Intervention	Prescription
Objective Number	Illinois Assessment Objectives Grade 4	SE/TE Lessons	Test Item(s)	Criterion Score	Student Score	SE/TE Lessons	Test Item(s)	Criterion Score	Student Score		
Goal 6: Number Sense											
6.4.03 (A)	Interpret whole numbers up to 100,000; demonstrate an understanding of the values of the digits and comparing and ordering the numbers.	1.1, 1.2, 2.1, 2.2	1, 4	2/2	___/2	1.1, 1.2, 2.1, 2.2	1, 4	2/2	___/2	**SK:** 5	**R, P, PS:** 1.1, 1.2, 2.1, 2.2 **EP:** SE pp. 14, Sets A–B; 34, Sets A–B **ATS:** TE pp. 2B, 4B, 20B, 22, 24B, 26 **MM:** FA/NLM Levels A–B, TNG/TT Levels A, I
6.4.04 (A)	Represent, order, and compare large numbers (up to 100,000) using various forms, including expanded notation (e.g., 853 = 8 × 100 + 5 × 10 + 3).	1.2, 2.1, 2.2	2	1/1	___/1	1.2, 2.1, 2.2	2	1/1	___/1	**SK:** 1	**R, P, PS:** 1.2, 2.1, 2.2 **EP:** SE pp. 14, Set B; 34, Sets A–B **ATS:** TE pp. 4B, 20B, 22, 24B, 26 **MM:** TNG/TT Level A, I; FA/NLM Level B
6.4.06 (A)	Exhibit an understanding of the base-ten number system by reading, naming, and writing decimals between 0 and 1 up through the hundredths.	26.2	16	1/1	___/1	26.2	16	1/1	___/1	**SK:** 36	**R, P, PS:** 26.2 **EP:** SE p. 578, Set B **ATS:** TE p. 564B
6.4.11 (D)	Select, use, and explain models to relate common fractions and mixed numbers (in halves, thirds, fourths, fifths, sixths, eighths and tenths); find equivalent fractions, mixed numbers, improper fractions, and decimals, and order fractions.	21.2, 21.3, 21.4, 21.5	11, 12	2/2	___/2	21.2, 21.3, 21.4, 21.5	11, 12	2/2	___/2	**SK:** 31–32	**R, P, PS:** 21.2, 21.3, 21.4, 21.5 **EP:** SE p. 462, Sets B–D **ATS:** TE pp. 448B, 450, 452B, 454, 456B, 458B, 460 **MM:** FA/FF Levels D–F; FA/NLM Levels E–G, I

Individual Record Form — **Practice for the ISAT Test**

Individual Record Form

Name _____ Date _____

PRETEST & POSTTEST

OBJECTIVES & LESSONS ASSESSED

Objective Number	Illinois Assessment Objectives Grade 4	SE/TE Lessons	PRETEST Test Item(s)	PRETEST Criterion Score	PRETEST Student Score	SE/TE Lessons	POSTTEST Test Item(s)	POSTTEST Criterion Score	POSTTEST Student Score	Intervention	Prescription
6.4.12 (D)	Identify and generate equivalent forms of common decimals and fractions less than one whole (halves, quarters, fifths, and tenths).	21.2, 26.3	14, 17	2/2	__/2	21.2, 26.3	14, 17	2/2	__/2	SK: 35	**R, P, PS:** 21.2, 26.3 **EP:** SE p. 462, Set B **ATS:** TE pp. 448B, 450, 566B **MM:** FA/FF Levels D–E, M; FA/NLM Level E
6.4.14 (B/C)	Add and subtract decimals through hundredths.	27.3, 27.4, 27.5	18, 19	2/2	__/2	27.3, 27.4, 27.5	18, 19	2/2	__/2	SK: 38	**R, P, PS:** 27.3, 27.4, 27.5 **EP:** SE p. 598, Sets C–E **ATS:** TE pp. 588B, 590B, 592B, 594 **MM:** TNG/TT Level L; TNG/BB Levels F–I
6.4.15 (B/C)	Make estimates appropriate to a given situation with whole numbers, fractions, and decimals by knowing when to estimate, and select the appropriate type of estimate including overestimate, underestimate, and range of estimate, and select the appropriate method of estimation.	3.2, 3.6	5, 10, 20	2/3	__/3	3.2, 3.6	5, 10, 20	2/3	__/3	SK: 7	**R, P, PS:** 3.2, 3.6 **EP:** SE p. 58, Set B **ATS:** TE pp. 44B, 46, 56B

Practice for the ISAT Test   xx   **Individual Record Form**

Individual Record Form

Name _____ Date _____

▲ PRETEST & POSTTEST

OBJECTIVES & LESSONS ASSESSED | PRETEST | POSTTEST

Objective Number	Illinois Assessment Objectives Grade 4	SE/TE Lessons	Test Item(s)	Criterion Score	Student Score	SE/TE Lessons	Test Item(s)	Criterion Score	Student Score	Intervention	Prescription
6.4.16 (B/C)	Compute with whole numbers: addition—up to three 3-digit numbers with regrouping, or two 4-digit numbers; subtraction—up to 3-digit numbers with regrouping; multiplication—up to 3-digit by 1-digit numbers with regrouping; division—up to 3-digit by 1-digit numbers with and without remainder.	8.2, 8.3, 8.4, 10.3, 10.4, 11.3, 12.1, 12.2, 13.1, 14.3, 14.4, 15.3	3, 7, 8, 9	3/4	__/4	8.2, 8.3, 8.4, 10.3, 10.4, 11.3, 12.1, 12.2, 13.1, 14.3, 14.4, 15.3	3, 7, 8, 9	3/4	__/4	**SK:** 10, 14–15, 24, 26	**R, P, PS:** 8.2, 8.3, 8.4, 10.3, 10.4, 11.3, 12.1, 12.2, 13.1, 14.3, 14.4, 15.3 **EP:** SE pp. 178, Sets B–C; 230, Sets C–D; 246, Set B; 264, Sets A–B; 290, Set A; 310, Sets C–D; 330, Set B **ATS:** TE pp. 166B, 168B, 170, 172B, 218B, 220, 222B, 224, 240B, 252B, 254, 256B, 278B, 300B, 302B, 304, 320B **MM:** TNG/UA Levels A, C–E, G–H, J–K, F, L, N–P, R
6.4.21 (B/C)	Use the inverse relationship of multiplication and division to compute and check results. Use these relationships to solve problems (e.g., 5 × 3 = 15 and 15 ÷ 3 = __).	15.5	13	1/1	__/1	15.5	13	1/1	__/1	**SK:** 26–27	**R, P, PS:** 15.5 **ATS:** TE p. 326B
6.4.24 (A)	Recognize equivalent representations of whole numbers and generate them by composing and decomposing numbers through the use of expanded notation to represent numbers (e.g., 3,206 = 3,000 + 200 + 6).	1.2	6	1/1	__/1	1.2	6	1/1	__/1	**SK:** 1	**R, P, PS:** 1.2 **EP:** SE p. 14, Set B **ATS:** TE p. 4B **MM:** TNG/TT Level A

Individual Record Form xxi **Practice for the ISAT Test**

Individual Record Form

Name _____ Date _____

PRETEST & POSTTEST

OBJECTIVES & LESSONS ASSESSED

Objective Number	Illinois Assessment Objectives Grade 4	SE/TE Lessons	PRETEST Test Item(s)	PRETEST Criterion Score	PRETEST Student Score	SE/TE Lessons	POSTTEST Test Item(s)	POSTTEST Criterion Score	POSTTEST Student Score	Intervention	Prescription
6.4.27 (A)	Divide regions or sets to represent a fraction; name and write the fractions represented by a given model (area/region, length/measurement, and set). (Fractions will include halves, thirds, fourths, and tenths.)	21.1, 21.4	15	1/1	__/1	21.1, 21.4	15	1/1	__/1	**SK:** 32	**R, P, PS:** 21.1, 21.4 **EP:** SE p. 484, Set A **ATS:** TE pp. 468B, 470, 478B **MM:** FA/FF Level G

Goal 7: Measurement

Objective Number	Illinois Assessment Objectives Grade 4	SE/TE Lessons	PRETEST Test Item(s)	PRETEST Criterion Score	PRETEST Student Score	SE/TE Lessons	POSTTEST Test Item(s)	POSTTEST Criterion Score	POSTTEST Student Score	Intervention	Prescription
7.4.02 (A/C)	Compute elapsed time in compound units (e.g., 1 hour and 30 minutes).	5.2	21	1/1	__/1	5.2	21	1/1	__/1	**SK:** 39	**R, P, PS:** 5.2 **EP:** SE p. 108, Set B **ATS:** TE pp. 98B, 100 **MM:** TNG/TT Level D
7.4.04 (A/C)	Solve problems that require a knowledge of the following units: inches—down to $\frac{1}{2}, \frac{1}{4}$ feet and ?; feet, yards, miles, millimeters, centimeters, meters, and kilometers; weight/mass—ounces, pounds, tons, grams, and kilograms; liquid volume—cups, pints, quarts, gallons, milliliters, and liters; area—square units; temperature (Celsius and Fahrenheit units).	24.4, 24.5, 25.1, 25.3, 25.4, 29.3, 29.4	25	1/1	__/1	24.4, 24.5, 25.1, 25.3, 25.4, 29.3, 29.4	25	1/1	__/1	**SK:** 41–42	**R, P, PS:** 24.4, 24.5, 25.1, 25.3, 25.4, 29.3, 29.4 **EP:** SE pp. 554, Set A; 638, Set B **ATS:** TE pp. 528B, 530B, 540B, 542, 546B, 548B, 634B, 636B **MM:** TNG/TT Levels N–O, Q; ISE/LL Levels I–J; ISE/PP Level S

Practice for the ISAT Test

Individual Record Form

Name _____ Date _____

▲ PRETEST & POSTTEST

OBJECTIVES & LESSONS ASSESSED | PRETEST | POSTTEST

Objective Number	Illinois Assessment Objectives Grade 4	SE/TE Lessons	Test Item(s)	Criterion Score	Student Score	SE/TE Lessons	Test Item(s)	Criterion Score	Student Score	Intervention	Prescription
7.4.05 (A/C) (OR)	Calculate the area and perimeter of a rectangle, triangle, or irregular shape using diagrams, models, and grids or by measuring. Use the appropriate units in the response [e.g., square centimeter (cm^2), square meter (m^2), square inch (in^2), or square yard (yd^2)].	28.2, 29.2	23, 30	2/2	__/2	28.2, 29.2	23, 30	2/2	__/2	SK: 44	R, P, PS: 28.2, 29.2 EP: SE pp. 622, Set A; 638, Set A ATS: TE pp. 614B, 616, 630B, 632 MM: ISE/PP Level R
7.4.06 (B)	Choose the appropriate units (metric and U.S.) to estimate the length, liquid volume, and weight/mass of given objects.	24.1	29	1/1	__/1	24.1	29	1/1	__/1	SK: 41	R, P, PS: 24.1 EP: SE p. 534, Set A ATS: TE pp. 518B, 520 MM: ISE/LL Level E
7.4.08 (B)	Estimate standard measurements of length, weight, and capacity.	24.4, 24.5, 25.1, 25.3, 25.4	26	1/1	__/1	24.4, 24.5, 25.1, 25.3, 25.4	26	1/1	__/1	SK: 41	R, P, PS: 24.4, 24.5, 25.1, 25.3, 25.4 EP: SE p. 554, Set A ATS: TE pp. 528B, 530B, 540B, 542, 546B, 548B MM: TNG/TT Levels N–O, Q; ISE/LL Levels I–J
7.4.09 (A/C)	Perform simple unit conversions within a system of measurement (e.g., feet to inches, yards to feet).	24.3, 25.2	27	1/1	__/1	24.3, 25.2	27	1/1	__/1	SK: 41	R, P, PS: 24.3, 25.2 EP: SE p. 534, Set C; 554, Set B ATS: TE p. 526B, 544B MM: TNG/TT Levels M, Q
7.4.10 (A/C)	Read temperature to the nearest degree from a Celsius thermometer and a Fahrenheit thermometer (does not require converting between °F and °C).	20.1, 20.2	22	1/1	__/1	20.1, 20.2	22	1/1	__/1	SK: 4	R, P, PS: 20.1, 20.2 EP: SE p. 432, Sets A–B ATS: TE pp. 420B, 422B MM: TNG/TT Level P

Individual Record Form — xxiii — **Practice for the ISAT Test**

Individual Record Form

Name _____ Date _____

PRETEST & POSTTEST

OBJECTIVES & LESSONS ASSESSED | PRETEST | POSTTEST

Objective Number	Illinois Assessment Objectives Grade 4	SE/TE Lessons	Test Item(s)	Criterion Score	Student Score	SE/TE Lessons	Test Item(s)	Criterion Score	Student Score	Intervention	Prescription
Goal 8: Algebra											
8.4.01 (A)	Know and extend a linear pattern by a well-defined rule or find a rule that fits the pattern (e.g., show a table that pairs number of horses with the number of legs calculated by counting by 4s or by multiplying the number of horses by 4).	4.4, 9.6	31, 34	2/2	__/2	4.4, 9.6	31, 34	2/2	__/2	SK: 55–56	**R, P, PS:** 4.4, 9.6 **EP:** SE pp. 82, Set D; 200, Set D **ATS:** TE pp. 74B, 198B **MM:** ISE/AA Level J–K
8.4.02 (A)	Identify or represent situations with well-defined patterns, such as rate of change, using words, tables, and graphs (e.g., represent in a bar graph the growth over five weeks of a plant that grows 1 inch per week).	7.1, 7.2, 7.3, 7.5	33	1/1	__/1	7.1, 7.2, 7.3, 7.5	33	1/1	__/1	SK: 63–64	**R, P, PS:** 7.1, 7.2, 7.3, 7.5 **EP:** SE p. 150, Sets A–B **ATS:** TE pp. 136B, 138B, 140B, 144B, 146 **MM:** TNG/AG Levels D, I–J
8.4.03 (C/D)	Determine values of variables in simple equations (e.g., $41 - y = 37$, $5 = m + 3$, and $c - 1 = 3$).	4.1, 4.3	28	1/1	__/1	4.1, 4.3	28	1/1	__/1	SK: 11, 54	**R, P, PS:** 4.1, 4.3 **EP:** SE p. 82, Sets A, C **ATS:** TE pp. 64B, 66, 70B, 72 **MM:** ISE/AA Levels G, S
8.4.05 (C/D)	Use parentheses to indicate which operation to perform first when writing expressions containing more than two terms and different operations.	9.2	35	1/1	__/1	9.2	35	1/1	__/1	SK: 56	**R, P, PS:** 9.2 **ATS:** TE p. 186B **MM:** ISE/AA Level Q
8.4.06 (C/D)	Interpret and evaluate mathematical expressions that use parentheses.	9.2, 9.3, 9.5	36	1/1	__/1	9.2, 9.3, 9.5	36	1/1	__/1	SK: 56	**R, P, PS:** 9.2, 9.3, 9.5 **EP:** SE p. 200, Sets B–C **ATS:** TE pp. 186B, 188B, 190, 194B, 196 **MM:** ISE/AA Levels I, Q

Individual Record Form

Name _____ Date _____

▲ PRETEST & POSTTEST

OBJECTIVES & LESSONS ASSESSED | PRETEST | POSTTEST

Objective Number	Illinois Assessment Objectives Grade 4	SE/TE Lessons	Test Item(s)	Criterion Score	Student Score	SE/TE Lessons	Test Item(s)	Criterion Score	Student Score	Intervention	Prescription
Goal 9: Geometry											
9.4.02 (A)	Identify regular and irregular polygons.	18.1	37	1/1	__/1	18.1	37	1/1	__/1	**SK:** 47	**R, P, PS:** 18.1 **EP:** SE p. 392, Set A **ATS:** TE p. 380B **MM:** ISE/PP Level D
9.4.07 (A)	Identify paths and movements using coordinate systems.	20.6	39	1/1	__/1	20.6	39	1/1	__/1	**SK:** 58	**R, P, PS:** 20.6 **EP:** SE p. 432, Set E **ATS:** TE p. 430B **MM:** TNG/AG Level G
9.4.08 (A)	Graph points and identify coordinates of points on the Cartesian coordinate plane (quadrant I only).	20.6	40	1/1	__/1	20.6	40	1/1	__/1	**SK:** 58	**R, P, PS:** 20.6 **EP:** SE p. 432, Set E **ATS:** TE p. 430B **MM:** TNG/AG Level G
9.4.09 (A)	Identify, describe, and classify polygons (including triangles, squares, rectangles, pentagons, hexagons, octagons).	18.1, 18.2, 18.3	38	1/1	__/1	18.1, 18.2, 18.3	38	1/1	__/1	**SK:** 47	**R, P, PS:** 18.1, 18.2, 18.3 **EP:** SE p. 392, Sets A–C **ATS:** TE pp. 380B, 382B, 384B, 386 **MM:** ISE/PP Levels D–E, G; TNG/TT Level H
Goal 10: Data Analysis, Statistics, and Probability											
10.4.01 (A/B)	Use information from a pictograph, bar graph, line graph, or a chart/table, with no more than two variables.	7.2, 7.6	41	1/1	__/1	7.2, 7.6	41	1/1	__/1	**SK:** 63–64	**R, P, PS:** 7.2, 7.6 **EP:** SE p. 150, Set A **ATS:** TE pp. 138B, 148B **MM:** TNG/AG Levels I–J

Individual Record Form   **Practice for the ISAT Test**

Individual Record Form

Name _____ Date _____

▲ PRETEST & POSTTEST

OBJECTIVES & LESSONS ASSESSED

Objective Number	Illinois Assessment Objectives Grade 4	PRETEST SE/TE Lessons	PRETEST Test Item(s)	PRETEST Criterion Score	PRETEST Student Score	POSTTEST SE/TE Lessons	POSTTEST Test Item(s)	POSTTEST Criterion Score	POSTTEST Student Score	Intervention	Prescription
10.4.04 (A/B)	Determine minimum value, maximum value, range, mode, and median for a data set with an odd number of data points.	6.2	42	1/1	—/1	6.2	42	1/1	—/1	SK: 62	**R, P, PS:** 6.2 **EP:** SE p. 130, Set B **ATS:** TE pp. 118B, 120
10.4.05 (C)	Classify events as certain, more likely, or less likely by experiments using objects such as counters, number cubes, spinners, or coins, where visual cues are unambiguous.	23.2	32	1/1	—/1	23.2	32	1/1	—/1	SK: 65–66	**R, P, PS:** 23.2 **EP:** SE p. 504, Set A **ATS:** TE p. 494B **MM:** FA/LC Levels D–E
10.4.06 (A/B)	Use information from a pictograph, bar graph, or a chart/table to answer questions about a situation (which assumes only one variable).	7.6	24	1/1	—/1	7.6	24	1/1	—/1	SK: 63–64	**R, P, PS:** 7.6 **ATS:** TE p. 148B

Extended-Response Items: Use rubric scoring

Objective Number	Illinois Assessment Objectives Grade 4	PRETEST SE/TE Lessons	PRETEST Test Item(s)	PRETEST Criterion Score	PRETEST Student Score	POSTTEST SE/TE Lessons	POSTTEST Test Item(s)	POSTTEST Criterion Score	POSTTEST Student Score	Intervention	Prescription
6.4.14 (B/C)	Add and subtract decimals through hundredths.	27.3, 27.4, 27.5	43			27.3, 27.4, 27.5	43			SK: 37–38	**R, P, PS:** 27.3, 27.4, 27.5 **EP:** SE p. 598, Sets C–E **ATS:** TE pp. 588B, 590B, 592B, 594 **MM:** TNG/TT Level L; TNG/BB Levels F–I
7.4.04 (A/C)	Solve problems that require a knowledge of the following units: inches—down to $\frac{1}{2}, \frac{1}{4}$; feet, yards, miles, mm, cm, m, km; weight/mass—ounces, pounds, tons, g, kg; liquid volume—cups, pints, qt, gal, mL, L; area—square units; temperature.	29.3, 29.4	44			29.3, 29.4	44			SK: 44	**R, P, PS:** 29.3, 29.4 **EP:** SE p. 638, Set B **ATS:** TE pp. 634B, 636B **MM:** ISE/PP Level S

Practice for the ISAT Test **Individual Record Form**

Individual Record Form

Name _____ Date _____

TEST UNIT 1

OBJECTIVES & LESSONS ASSESSED

Objective Number	Illinois Assessment Objectives Grade 4	SE/TE Lessons	Test Item(s)	Criterion Score	Student Score	Intervention	Prescription
Goal 6: Number Sense							
6.4.02 (A)	Order and compare whole numbers and decimals to two decimal places.	2.1, 2.2	1, 2	2/2	__/2	**SK:** 5	**R, P, PS:** 2.1, 2.2 **EP:** SE p. 34, Sets A–B **ATS:** TE pp. 20B, 22, 24B, 26 **MM:** TNG/TT Level I; FA/NLM Level B
6.4.03 (A)	Interpret whole numbers up to 100,000; demonstrate an understanding of the values of the digits and comparing and ordering the numbers.	1.1, 1.2, 2.1, 2.2	3, 4	2/2	__/2	**SK:** 1, 5	**R, P, PS:** 1.1, 1.2, 2.1, 2.2 **EP:** SE pp. 14, Sets A–B; 34, Sets A–B **ATS:** TE pp. 2B, 4B, 20B, 22, 24B, 26 **MM:** TNG/TT Levels A, I; FA/NLM Levels A–B
6.4.04 (A)	Represent, order, and compare large numbers (up to 100,000) using various forms, including expanded notation (e.g., $853 = 8 \times 100 + 5 \times 10 + 3$).	1.2, 2.1, 2.2	5	1/1	__/1	**SK:** 1, 5	**R, P, PS:** 1.2, 2.1, 2.2 **EP:** SE pp. 14, Set B; 34, Sets A–B **ATS:** TE pp. 4B, 20B, 22, 24B, 26 **MM:** TNG/TT Levels A, I; FA/NLM Level B
6.4.15 (B/C)	Make estimates appropriate to a given situation with whole numbers, fractions, and decimals by knowing when to estimate, and select the appropriate type of estimate including overestimate, underestimate, and range of estimate, and select the appropriate method of estimation.	3.2, 3.6	7, 8, 12	2/3	__/3	**SK:** 8, 10	**R, P, PS:** 3.2, 3.6 **EP:** SE p. 58, Set B **ATS:** TE pp. 44B, 46, 56B

Individual Record Form

Name _____ Date _____

▲ **TEST UNIT 1** *(continued)*

OBJECTIVES & LESSONS ASSESSED

Objective Number	Illinois Assessment Objectives Grade 4	SE/TE Lessons	Test Item(s)	Criterion Score	Student Score	Intervention	Prescription
6.4.16 (B/C)	Compute with whole numbers: addition—up to three 3-digit numbers with regrouping, or two 4-digit numbers; subtraction—up to 3-digit numbers with regrouping; multiplication—up to 3-digit by 1-digit numbers with regrouping; division—up to 3-digit by 1-digit numbers with and without remainder.	3.3, 3.4	6	1/1	__/1	SK: 8, 10	**R, P, PS:** 3.3, 3.4 **EP:** SE p. 58, Sets C–D **ATS:** TE pp. 48B, 50B **MM:** TNG/TT Levels B–C
6.4.17 (B/C)	Round whole numbers through the millions to the nearest ten, hundred, thousand, ten thousand, or hundred thousand in contextual problems.	2.4	9	1/1	__/1	SK: 6	**R, P, PS:** 2.4 **EP:** SE p. 34, Set C **ATS:** TE pp. 30B, 32 **MM:** FA/NLM Level C
6.4.19 (B/C)	Use estimation to verify the reasonableness of calculated results.	3.6	10	1/1	__/1	SK: 7	**R, P, PS:** 3.6 **ATS:** TE p. 56B
6.4.20 (A)	Terminology: Know that in $q = x \div y$, q is the quotient; in $x + y = s$, s is the sum; in $x - y = d$, d is the difference; in $(x)(y) = p$, p is the product.	3.2	13	1/1	__/1	SK: 8, 10	**R, P, PS:** 3.2 **EP:** SE p. 58, Set B **ATS:** TE pp. 44B, 46
6.4.22 (A)	Recognize and write numerals from dictation up to 9,999.	1.2	11	1/1	__/1	SK: 1	**R, P, PS:** 1.2 **EP:** SE p. 14, Set B **ATS:** TE p. 4B **MM:** TNG/TT Level A

Individual Record Form

Name _____   Date _____

▶ **TEST UNIT 1** *(continued)*

OBJECTIVES & LESSONS ASSESSED

Objective Number	Illinois Assessment Objectives Grade 4	SE/TE Lessons	Test Item(s)	Criterion Score	Student Score	Intervention	Prescription
6.4.23 (A)	Order whole numbers up to 999.	2.2	14	1/1	__/1	SK: 5	**R, P, PS:** 2.2 **EP:** SE p. 34, Set B **ATS:** TE pp. 24B, 26
6.4.24 (A)	Recognize equivalent representations of whole numbers and generate them by composing and decomposing numbers through the use of expanded notation to represent numbers (e.g. 3,206 = 3,000 + 200 + 6).	1.2	15, 16	2/2	__/2	SK: 1	**R, P, PS:** 1.2 **EP:** SE p. 14, Set B **ATS:** TE p. 4B **MM:** TNG/TT Level A
6.4.25 (A)	Represent and order whole numbers between 0 and 9,999, using symbols (>, <, or =) and words (greater than, less than, or equal to).	2.1	17	1/1	__/1	SK: 5	**R, P, PS:** 2.1 **EP:** SE p. 34, Set A **ATS:** TE pp. 20B, 22 **MM:** TNG/TT Level I; FA/NLM Level B
6.4.28 (B/C)	Solve addition and subtraction number sentences and word problems with numbers up to 3 digits.	4.3, 4.5	18, 19	2/2	__/2	SK: 11, 54	**R, P, PS:** 4.3, 4.5 **EP:** SE p. 82, Sets C, E **ATS:** TE pp. 70B, 72, 76B, 78 **MM:** ISE/AA Level S
Goal 8: Algebra							
8.4.01 (A)	Know and extend a linear pattern by a well-defined rule or find a rule that fits the pattern (e.g., show a table that pairs number of horses with the number of legs calculated by counting by 4s or by multiplying the number of horses by 4).	4.4	22, 23, 24	2/3	__/3	SK: 55	**R, P, PS:** 4.4 **EP:** SE p. 82, Set D **ATS:** TE p. 74B **MM:** ISE/AA Level J

Individual Record Form   xxix   **Practice for the ISAT Test**

Individual Record Form

Name _____ Date _____

▶ **TEST UNIT 1** *(continued)*

OBJECTIVES & LESSONS ASSESSED

Objective Number	Illinois Assessment Objectives Grade 4	SE/TE Lessons	Test Item(s)	Criterion Score	Student Score	Intervention	Prescription
8.4.03 (C/D)	Determine values of variables in simple equations (e.g., $41 - y = 37$, $5 = m + 3$, and $c - 1 = 3$).	4.1, 4.3	20	1/1	___/1	SK: 11, 54	**R, P, PS:** 4.1, 4.3 **EP:** SE p. 82, Sets A, C **ATS:** TE pp. 64B, 66, 70B, 72 **MM:** ISE/AA Levels G, S
8.4.07 (A)	Demonstrate an understanding of equality by recognizing that "=" links equivalent quantities (e.g., $4 \times 3 = 2 \times 6$).	4.5	21	1/1	___/1	SK: 11	**R, P, PS:** 4.5 **EP:** SE p. 82, Set E **ATS:** TE pp. 76B, 78

Extended-Response Items: Use rubric scoring

6.4.04 (A)	Represent, order, and compare large numbers (up to 100,000) using various forms, including expanded notation (e.g., $853 = 8 \times 100 + 5 \times 10 + 3$).	1.2, 2.1, 2.2	25			SK: 1, 5	**R, P, PS:** 1.2, 2.1, 2.2 **EP:** SE pp. 14, Set B; 34, Sets A–B **ATS:** TE pp. 4B, 20B, 22, 24B, 26 **MM:** TNG/TT Levels A, I; FA/NLM Level B
6.4.28 (B/C)	Solve addition and subtraction number sentences and word problems with numbers up to 3 digits.	4.3, 4.5	26			PS: 4	**R, P, PS:** 4.3, 4.5 **EP:** SE p. 82, Sets C, E **ATS:** TE pp. 70B, 72, 76B, 78 **MM:** ISE/AA Level S

Practice for the ISAT Test   xxx   **Individual Record Form**

Individual Record Form

Name _____ Date _____

▶ **TEST UNIT 2**

OBJECTIVES & LESSONS ASSESSED

Objective Number	Illinois Assessment Objectives Grade 4	SE/TE Lessons	Test Item(s)	Criterion Score	Student Score	Intervention	Prescription
Goal 7: Measurement							
7.4.02 (A/C)	Compute elapsed time in compound units (e.g., 1 hour and 30 minutes).	5.2	1, 2, 3	2/3	__/3	**SK:** 39	**R, P, PS:** 5.2 **EP:** SE p. 108, Set B **ATS:** TE pp. 98B, 100 **MM:** TNG/TT Level D
Goal 8: Algebra							
8.4.02 (A)	Identify or represent situations with well-defined patterns, such as rate of change, using words, tables, and graphs (e.g., represent in a bar graph the growth over five weeks of a plant that grows 1 inch per week).	7.1, 7.2, 7.3, 7.5	10	1/1	__/1	**SK:** 63–64	**R, P, PS:** 7.1, 7.2, 7.3, 7.5 **EP:** SE pp. 150, Sets A–B **ATS:** TE pp. 136B, 138B, 140B, 144B, 146 **MM:** TNG/AG Levels D, I–J
Goal 10: Data Analysis, Statistics, and Probability							
10.4.01 (A/B)	Use information from a pictograph, bar graph, line graph, or a chart/table, with no more than two variables.	7.2, 7.6	5, 8, 9, 11, 19	3/5	__/5	**SK:** 63–64	**R, P, PS:** 7.2, 7.6 **EP:** SE p. 150, Set A **ATS:** TE pp. 138B, 148B **MM:** TNG/AG Levels I–J
10.4.02 (A/B)	Match a data set to a graph and vice versa.	6.1, 6.4, 6.6, 7.1, 7.3, 7.4	4, 6, 7	2/3	__/3	**SK:** 61–64	**R, P, PS:** 6.1, 6.4, 6.6, 7.1, 7.3, 7.4 **EP:** SE p. 130, Sets A, D **ATS:** TE pp. 114B, 116, 124B, 128B, 136B, 140B, 142B **MM:** TNG/AG Level D
10.4.03 (A/B)	Identify different representations of the same data.	6.5	14	1/1	__/1	**SK:** 61–62	**R, P, PS:** 6.5 **EP:** SE p. 130, Set E **ATS:** TE p. 126B
10.4.04 (A/B)	Determine minimum value, maximum value, range, mode, and median for a data set with an odd number of data points.	6.2	12, 13, 15	2/3	__/3	**SK:** 62	**R, P, PS:** 6.2 **EP:** SE p. 130, Set B **ATS:** TE pp. 118B, 120

Individual Record Form

Name _____ Date _____

▶ **TEST UNIT 2** *(continued)*

OBJECTIVES & LESSONS ASSESSED

Objective Number	Illinois Assessment Objectives Grade 4	SE/TE Lessons	Test Item(s)	Criterion Score	Student Score	Intervention	Prescription
10.4.06 (A/B)	Use information from a pictograph, bar graph, or a chart/table to answer questions about a situation (which assumes only one variable).	7.6	16, 17, 18, 20	3/4	__/4	**SK:** 61, 63–64	**R, P, PS:** 7.6 **ATS:** TE p. 148B
Extended-Response Items: Use rubric scoring							
10.4.01 (A/B)	Use information from a pictograph, bar graph, line graph, or a chart/table, with no more than two variables.	7.2, 7.6	21			**SK:** 61, 63–64	**R, P, PS:** 7.2, 7.6 **EP:** SE p. 150, Set A **ATS:** TE pp. 138B, 148B **MM:** TNG/AG Levels I–J
10.4.06 (A/B)	Use information from a pictograph, bar graph, or a chart/table to answer questions about a situation (which assumes only one variable).	7.6	22			**SK:** 61, 63–64	**R, P, PS:** 7.6 **ATS:** TE p. 148B

Practice for the ISAT Test

Individual Record Form

Name _____ Date _____

▲ TEST UNIT 3

OBJECTIVES & LESSONS ASSESSED

Objective Number	Illinois Assessment Objectives Grade 4	SE/TE Lessons	Test Item(s)	Criterion Score	Student Score	Intervention	Prescription
Goal 6: Number Sense							
6.4.16 (B/C)	Compute with whole numbers: addition—up to three 3-digit numbers with regrouping, or two 4-digit numbers; subtraction—up to 3-digit numbers with regrouping; multiplication—up to 3-digit by 1-digit numbers with regrouping; division—up to 3-digit by 1-digit numbers with and without remainder.	8.2, 8.3, 8.4	1, 2, 3, 19, 20	3/5	___/5	SK: 12, 21	**R, P, PS:** 8.2, 8.3, 8.4 **EP:** SE p. 178, Sets B–C **ATS:** TE pp. 166B, 168B, 170, 172B **MM:** TNG/UA Levels A, C–E, G–H
6.4.21 (B/C)	Use the inverse relationship of multiplication and division to compute and check results. Use these relationships to solve problems (e.g., 5 × 3 = 15 and 15 ÷ 3 = ___).	8.1, 8.3	4, 5	2/2	___/2	SK: 12, 21	**R, P, PS:** 8.1, 8.3 **EP:** SE p. 178, Sets A, C **ATS:** TE pp. 164B, 168B, 170 **MM:** TNG/UA Levels C, G
6.4.26 (B/C)	Use whole number multiplication and division (know the multiplication tables through 12 × 12).	8.2, 8.3, 8.4	6, 7, 8	2/3	___/3	SK: 12, 21	**R, P, PS:** 8.2, 8.3, 8.4 **EP:** SE p. 178, Sets B–C **ATS:** TE pp. 166B, 168B, 170, 172B **MM:** TNG/UA Levels A, C–E, G–H
Goal 8: Algebra							
8.4.01 (A)	Know and extend a linear pattern by a well-defined rule or find a rule that fits the pattern (e.g., show a table that pairs number of horses with the number of legs calculated by counting by 4s or by multiplying the number of horses by 4).	9.6	9, 10, 17	2/3	___/3	PS: 9	**R, P, PS:** 9.6 **EP:** SE p. 200, Set D **ATS:** TE p. 198B **MM:** ISE/AA Level K

Individual Record Form   xxxiii   **Practice for the ISAT Test**

Individual Record Form

Name _____ Date _____

TEST UNIT 3 (continued)

OBJECTIVES & LESSONS ASSESSED

Objective Number	Illinois Assessment Objectives Grade 4	SE/TE Lessons	Test Item(s)	Criterion Score	Student Score	Intervention	Prescription
8.4.03 (C/D)	Determine values of variables in simple equations (e.g., $41 - y = 37$, $5 = m + 3$, and $c - 1 = 3$).	9.3	11, 12	2/2	__/2	SK: 56	**R, P, PS:** 9.3 **EP:** SE p. 200, Set B **ATS:** TE pp. 188B, 190 **MM:** ISE/AA Level I
8.4.05 (C/D)	Use parentheses to indicate which operation to perform first when writing expressions containing more than two terms and different operations.	9.2	13, 14	2/2	__/2	SK: 56	**R, P, PS:** 9.2 **ATS:** TE p. 186B **MM:** ISE/AA Level Q
8.4.06 (C/D)	Interpret and evaluate mathematical expressions that use parentheses.	9.2, 9.3, 9.5	15, 16	2/2	__/2	SK: 56	**R, P, PS:** 9.2, 9.3, 9.5 **EP:** SE p. 200, Sets B–C **ATS:** TE pp. 186B, 188B, 190, 194B, 196 **MM:** ISE/AA Levels I, Q
8.4.07 (A)	Demonstrate an understanding of equality by recognizing that "=" links equivalent quantities (e.g., $4 \times 3 = 2 \times 6$).	9.5	18	1/1	__/1	SK: 12	**R, P, PS:** 9.5 **EP:** SE p. 200, Set C **ATS:** TE pp. 194B, 196

Individual Record Form

Name _____   Date _____

▶ **TEST UNIT 3** *(continued)*

OBJECTIVES & LESSONS ASSESSED

Objective Number	Illinois Assessment Objectives Grade 4	SE/TE Lessons	Test Item(s)	Criterion Score	Student Score	Intervention	Prescription
Extended-Response Items: Use rubric scoring							
6.4.21 (B/C)	Use the inverse relationship of multiplication and division to compute and check results. Use these relationships to solve problems (e.g., 5 × 3 = 15 and 15 ÷ 3 = ___).	8.1, 8.3	21			**SK:** 13, 22–23	**R, P, PS:** 8.1, 8.3 **EP:** SE p. 178, Sets A, C **ATS:** TE pp. 164B, 168B, 170 **MM:** TNG/UA Levels C, G
8.4.05 (C/D)	Use parentheses to indicate which operation to perform first when writing expressions containing more than two terms and different operations.	9.2	22			**SK:** 56	**R, P, PS:** 9.2 **ATS:** TE p. 186B **MM:** ISE/AA Level Q

Individual Record Form   xxxv   **Practice for the ISAT Test**

Individual Record Form

Name _____ Date _____

TEST UNIT 4

OBJECTIVES & LESSONS ASSESSED

Objective Number	Illinois Assessment Objectives Grade 4	SE/TE Lessons	Test Item(s)	Criterion Score	Student Score	Intervention	Prescription
Goal 6: Number Sense							
6.4.15 (B/C)	Make estimates appropriate to a given situation with whole numbers, fractions, and decimals by knowing when to estimate, and select the appropriate type of estimate including overestimate, underestimate, and range of estimate, and select the appropriate method of estimation.	10.2, 11.4	1, 2, 3, 4, 5	3/5	__/5	**PS:** 11	**R, P, PS:** 10.2, 11.4 **EP:** SE pp. 230, Set B; 246, Set C **ATS:** TE pp. 216B, 242B
6.4.16 (B/C)	Compute with whole numbers: addition—up to three 3-digit numbers with regrouping, or two 4-digit numbers; subtraction—up to 3-digit numbers with regrouping; multiplication—up to 3-digit by 1-digit numbers with regrouping; division—up to 3-digit by 1-digit numbers with and without remainder.	10.3, 10.4, 11.3, 12.1, 12.2	6, 8, 10, 18, 19	3/5	__/5	**SK:** 14–15, 23	**R, P, PS:** 10.3, 10.4, 11.3, 12.1, 12.2 **EP:** SE pp. 230, Sets C–D; 246, Set B; 264, Sets A–B **ATS:** TE pp. 218B, 220, 222B, 224, 240B, 252B, 254, 256B **MM:** TNG/UA Levels J–K, F
6.4.26 (B/C)	Use whole number multiplication and division (know the multiplication tables through 12 × 12).	11.3	11, 12, 13, 14, 15	3/5	__/5	**SK:** 15	**R, P, PS:** 11.3 **EP:** SE p. 246, Set B **ATS:** TE p. 240B

Practice for the ISAT Test — xxxvi — Individual Record Form

Individual Record Form

Name _____ Date _____

TEST UNIT 4 *(continued)*

OBJECTIVES & LESSONS ASSESSED

Objective Number	Illinois Assessment Objectives Grade 4	SE/TE Lessons	Test Item(s)	Criterion Score	Student Score	Intervention	Prescription
Goal 8: Algebra							
8.4.03 (C/D)	Determine values of variables in simple equations (e.g., $41 - y = 37$, $5 = m + 3$, and $c - 1 = 3$).	4.3	7, 9	2/2	__/2	**SK:** 11, 54	**R, P, PS:** 4.3 **EP:** SE p. 82, Set C **ATS:** TE pp. 70B, 72 **MM:** ISE/AA Level S
8.4.04 (A)	Solve simple problems concerning the functional relationship between two quantities (e.g., calculate the total cost of several items given the unit cost).	11.5, 12.5	16, 17, 20	2/3	__/3	**PS:** 11	**R, P, PS:** 11.5, 12.5 **ATS:** TE pp. 244B, 262B **MM:** TNG/BB Levels K–L
Extended-Response Items: Use rubric scoring							
6.4.16 (B/C)	Compute with whole numbers: addition—up to three 3-digit numbers with regrouping, or two 4-digit numbers; subtraction—up to 3-digit numbers with regrouping; multiplication—up to 3-digit by 1-digit numbers with regrouping; division—up to 3-digit by 1-digit numbers with and without remainder.	10.3, 10.4, 11.3, 12.1, 12.2	21, 22			**SK:** 15	**R, P, PS:** 10.3, 10.4, 11.3, 12.1, 12.2 **EP:** SE pp. 230, Sets C–D; 246, Set B; 264, Sets A–B **ATS:** TE pp. 218B, 220, 222B, 224, 240B, 252B, 254, 256B **MM:** TNG/UA Levels J–K, F

Individual Record Form   xxxvii   Practice for the ISAT Test

Individual Record Form

Name _____ Date _____

TEST UNIT 5

OBJECTIVES & LESSONS ASSESSED

Objective Number	Illinois Assessment Objectives Grade 4	SE/TE Lessons	Test Item(s)	Criterion Score	Student Score	Intervention	Prescription
Goal 6: Number Sense							
6.4.07 (A)	Perform prime factorization of all whole numbers through 20.	16.3	1, 3, 4	2/3	___/3	SK: 12–13, 23	R, P, PS: 16.3 EP: SE p. 350, Set C ATS: TE pp. 342B, 344
6.4.08 (A)	Identify all prime numbers through 20.	16.3	2, 5, 6	2/3	___/3	SK: 12–13, 23	R, P, PS: 16.3 EP: SE p. 350, Set C ATS: TE pp. 342B, 344
6.4.09 (A)	Identify classes (i.e., odds/evens, factors/multiples, squares) to which a number may belong, and identify the numbers in those classes. Use these in the solution of problems.	16.2, 16.5	7, 8, 10	2/3	___/3	PS: 16	R, P, PS: 16.2, 16.5 EP: SE p. 350, Sets B, D ATS: TE pp. 338B, 340, 348B MM: ISE/AA Level P
6.4.15 (B/C)	Make estimates appropriate to a given situation with whole numbers, fractions, and decimals by knowing when to estimate, and select the appropriate type of estimate including overestimate, underestimate, and range of estimate, and select the appropriate method of estimation.	14.1	9, 11, 12, 14	3/4	___/4	PS: 14	R, P, PS: 14.1 EP: SE p. 310, Set A ATS: TE p. 296B
6.4.16 (B/C)	Compute with whole numbers: addition—up to three 3-digit numbers with regrouping, or two 4-digit numbers; subtraction—up to 3-digit numbers with regrouping; multiplication—up to 3-digit by 1-digit numbers with regrouping; division—up to 3-digit by 1-digit numbers with and without remainder.	13.1, 14.3, 14.4, 15.3	13, 15, 17, 18	3/4	___/4	SK: 10, 23–24, 26–27	R, P, PS: 13.1, 14.3, 14.4, 15.3 EP: SE pp. 290, Set A; 310, Sets C–D; 330, Set B ATS: TE pp. 278B, 300B, 302B, 304, 320B MM: TNG/UA Levels L, N–P, R

Practice for the ISAT Test

Individual Record Form

Name _____   Date _____

▲ **TEST UNIT 5** *(continued)*

OBJECTIVES & LESSONS ASSESSED

Objective Number	Illinois Assessment Objectives Grade 4	SE/TE Lessons	Test Item(s)	Criterion Score	Student Score	Intervention	Prescription
6.4.21 (B/C)	Use the inverse relationship of multiplication and division to compute and check results. Use these relationships to solve problems (e.g., 5 × 3 = 15 and 15 ÷ 3 = ___).	15.5	16, 19, 20, 21	3/4	___/4	**SK:** 26	**R, P, PS:** 15.5 **ATS:** TE p. 326B

Extended-Response Items: Use rubric scoring

6.4.16 (B/C)	Compute with whole numbers: addition—up to three 3-digit numbers with regrouping, or two 4-digit numbers; subtraction—up to 3-digit numbers with regrouping; multiplication—up to 3-digit by 1-digit numbers with regrouping; division—up to 3-digit by 1-digit numbers with and without remainder.	13.1, 14.3, 14.4, 15.3	22, 23			**SK:** 10, 23	**R, P, PS:** 13.1, 14.3, 14.4, 15.3 **EP:** SE pp. 290, Set A; 310, Sets C–D; 330, Set B **ATS:** TE pp. 278B, 300B, 302B, 304, 320B **MM:** TNG/UA Levels L, N–P, R

Individual Record Form   xxxix   **Practice for the ISAT Test**

Individual Record Form

Name _____ Date _____

TEST UNIT 6

OBJECTIVES & LESSONS ASSESSED

Goal 7: Measurement

Objective Number	Illinois Assessment Objectives Grade 4	SE/TE Lessons	Test Item(s)	Criterion Score	Student Score	Intervention	Prescription
7.4.04 (A/C)	Solve problems that require a knowledge of the following units: inches—down to $\frac{1}{2}$, $\frac{1}{4}$ feet and ?; feet, yards, miles, millimeters, centimeters, meters, and kilometers; weight/mass—ounces, pounds, tons, grams, and kilograms; liquid volume—cups, pints, quarts, gallons, milliliters, and liters; area—square units; temperature (Celsius and Fahrenheit units).	20.1, 20.2	3	1/1	__/1	SK: 4	R, P, PS: 20.1, 20.2 EP: SE p. 432, Sets A–B ATS: TE pp. 420B, 422B MM: TNG/TT Level P
7.4.10 (A/C)	Read temperature to the nearest degree from a Celsius thermometer and a Fahrenheit thermometer (does not require converting between °F and °C).	20.1, 20.2	1, 2	2/2	__/2	SK: 4	R, P, PS: 20.1, 20.2 EP: SE p. 432, Sets A–B ATS: TE pp. 420B, 422B MM: TNG/TT Level P

Goal 9: Geometry

Objective Number	Illinois Assessment Objectives Grade 4	SE/TE Lessons	Test Item(s)	Criterion Score	Student Score	Intervention	Prescription
9.4.02 (A)	Identify regular and irregular polygons.	18.1	4, 5, 9, 13	3/4	__/4	SK: 48	R, P, PS: 18.1 EP: SE p. 392, Set A ATS: TE p. 380B MM: ISE/PP Level D
9.4.03 (A)	Identify the parts of a circle (radius, diameter, and circumference).	18.5	7, 8	2/2	__/2	SK: 47	R, P, PS: 18.5 EP: SE p. 392, Set D ATS: TE p. 390B
9.4.04 (B)	Differentiate between polygons and non-polygons.	18.1	6	1/1	__/1	SK: 47	R, P, PS: p. 18.1 EP: SE p. 392, Set A ATS: TE p. 380B MM: ISE/PP Level D

Practice for the ISAT Test

Individual Record Form

Name _____ Date _____

TEST UNIT 6 (continued)

OBJECTIVES & LESSONS ASSESSED

Objective Number	Illinois Assessment Objectives Grade 4	SE/TE Lessons	Test Item(s)	Criterion Score	Student Score	Intervention	Prescription
9.4.07 (A)	Identify paths and movements using coordinate systems.	20.6	10	1/1	__/1	SK: 58	R, P, PS: 20.6 EP: SE p. 432, Set E ATS: TE p. 430B MM: TNG/AG Level G
9.4.08 (A)	Graph points and identify coordinates of points on the Cartesian coordinate plane (quadrant I only).	20.6	11	1/1	__/1	SK: 58	R, P, PS: 20.6 EP: SE p. 432, Set E ATS: TE p. 430B MM: TNG/AG Level G
9.4.09 (A)	Identify, describe, and classify polygons (including triangles, squares, rectangles, pentagons, hexagons, and octagons).	18.1, 18.2, 18.3	14, 15	2/2	__/2	SK: 47–48	R, P, PS: 18.1, 18.2, 18.3 EP: SE p. 392, Sets A–C ATS: TE pp. 380B, 382B, 384B, 386 MM: ISE/PP Levels D–E, G; TNG/TT Level H
9.4.10 (A)	Determine the distance between two points on the number line in whole numbers.	16.3	12	1/1	__/1	PS: 16	R, P, PS: 16.3 ATS: TE p. 281B MM: CCD/WW Level I
Extended-Response Items: Use rubric scoring
9.4.02 (A)	Identify regular and irregular polygons.	18.1	16			SK: 48	R, P, PS: 18.1 EP: SE p. 392, Set A ATS: TE p. 380B MM: ISE/PP Level D
9.4.08 (A)	Graph points and identify coordinates of points on the Cartesian coordinate plane (quadrant I only).	20.6	17			SK: 58	R, P, PS: 20.6 EP: SE p. 432, Set E ATS: TE p. 430B MM: TNG/AG Level G

Individual Record Form   xli   **Practice for the ISAT Test**

Individual Record Form

Name _____ Date _____

TEST UNIT 7

OBJECTIVES & LESSONS ASSESSED

Objective Number	Illinois Assessment Objectives Grade 4	SE/TE Lessons	Test Item(s)	Criterion Score	Student Score	Intervention	Prescription
Goal 6: Number Sense							
6.4.01 (A)	Compare the numerical value of two fractions having like and unlike denominators up to twelfths, using concrete or pictorial models involving areas/regions, lengths/measurements, and sets.	21.3	1, 2	2/2	__/2	SK: 31	**R, P, PS:** 21.3 **EP:** SE p. 462, Set C **ATS:** TE pp. 452B, 454 **MM:** FA/FF Level F; FA/NLM Levels G, I
6.4.05 (A)	Identify on a number line the relative position of positive fractions, positive mixed numbers, and positive decimals to two decimal places.	21.1, 21.5	3, 4	2/2	__/2	PS: 21	**R, P, PS:** 21.1, 21.5 **EP:** SE p. 462, Sets A, D **ATS:** TE pp. 446B, 458B, 460 **MM:** FA/FF Levels B–C; FA/NLM Level F
6.4.11 (D)	Select, use, and explain models to relate common fractions and mixed numbers (in halves, thirds, fourths, fifths, sixths, eighths, tenths); find equivalent fractions, mixed numbers, improper fractions, and decimals, and order fractions.	21.2, 21.3, 21.4, 21.5	5, 6, 7, 8, 9	3/5	__/5	SK: 31–32	**R, P, PS:** 21.2, 21.3, 21.4, 21.5 **EP:** SE p. 462, Sets B–D **ATS:** TE pp. 448B, 450, 452B, 454, 456B, 458B, 460 **MM:** FA/FF Levels D–F; FA/NLM Levels E–G, I
6.4.13 (B/C)	Add and subtract fractions with like denominators in simple computations or in word problems.	22.1, 22.2	10, 11	2/2	__/2	PS: 22	**R, P, PS:** 22.1, 22.2 **EP:** SE p. 484, Set A **ATS:** TE pp. 468B, 470, 472B **MM:** FA/FF Levels G–H

Individual Record Form

Name _____   Date _____

TEST UNIT 7 (continued)

OBJECTIVES & LESSONS ASSESSED

Objective Number	Illinois Assessment Objectives Grade 4	SE/TE Lessons	Test Item(s)	Criterion Score	Student Score	Intervention	Prescription
6.4.18 (A)	Establish benchmarks (well known numbers used as meaningful points of comparison) for whole numbers, decimals, and fractions (e.g., $\frac{1}{2} = 0.5$, $0.25 = \frac{1}{4}$).	22.1	12	1/1	__/1	SK: 33	R, P, PS: 22.1 EP: SE p. 484, Set A ATS: TE pp. 468B, 470 MM: FA/FF Level G
6.4.27 (A)	Divide regions or sets to represent a fraction; name and write the fractions represented by a given model (area/region, length/measurement, and set). (Fractions will include halves, thirds, fourths, tenths).	21.1, 21.4	13, 14	2/2	__/2	SK: 31–32	R, P, PS: 21.1, 21.4 EP: SE p. 484, Set A ATS: TE pp. 468B, 470, 478B MM: FA/FF Level G
Goal 10: Data Analysis, Statistics, and Probability							
10.4.05 (C)	Classify events as certain, more likely, or less likely by experiments using objects such as counters, number cubes, spinners, or coins, where visual cues are unambiguous.	23.2	15, 16, 17	2/3	__/3	SK: 66	R, P, PS: 23.2 EP: SE p. 504, Set A ATS: TE p. 494B MM: FA/LC Levels D–E
Extended-Response Items: Use rubric scoring							
6.4.05 (A)	Identify on a number line the relative position of positive fractions, positive mixed numbers, and positive decimals to two decimal places.	21.1, 21.5	18			SK: 31–32	R, P, PS: 21.1, 21.5 EP: SE p. 462, Sets A, D ATS: TE pp. 446B, 458B, 460 MM: FA/FF Levels B–C; FA/NLM Level F
6.4.27 (A)	Divide regions or sets to represent a fraction; name and write the fractions represented by a given model (area/region, length/measurement, and set).	21.1, 21.4	19			SK: 32	R, P, PS: 21.1, 21.4 EP: SE p. 484, Set A ATS: TE pp. 468B, 470, 478B MM: FA/FF Level G

Individual Record Form

Name _____ Date _____

TEST UNIT 8

OBJECTIVES & LESSONS ASSESSED

Goal 6: Number Sense

Objective Number	Illinois Assessment Objectives Grade 4	SE/TE Lessons	Test Item(s)	Criterion Score	Student Score	Intervention	Prescription
6.4.02 (A)	Order and compare whole numbers and decimals to two decimal places.	26.5	1, 2	2/2	__/2	**SK:** 35–36	**R, P, PS:** 26.5 **EP:** SE p. 578, Set D **ATS:** TE pp. 572B, 574 **MM:** FA/NLM Levels P–Q
6.4.05 (A)	Identify on a number line the relative position of positive fractions, positive mixed numbers, and positive decimals to two decimal places.	26.1, 26.2	3, 4	2/2	__/2	**SK:** 35–36	**R, P, PS:** 26.1, 26.2 **EP:** SE p. 578, Sets A–B **ATS:** TE pp. 560B, 562, 564B **MM:** FA/NLM Levels M–N; FA/FF Levels L, N
6.4.06 (A)	Exhibit an understanding of the base-ten number system by reading, naming, and writing decimals between 0 and 1 up through the hundredths.	26.2	5	1/1	__/1	**SK:** 35–36	**R, P, PS:** 26.2 **EP:** SE p. 578, Set B **ATS:** TE p. 564B
6.4.10 (A)	Recognize equivalent representations for decimals and generate them by composing and decomposing numbers (e.g., 0.15 = 0.1 + 0.05).	26.2, 26.3	6	1/1	__/1	**PS:** 26	**R, P, PS:** 26.2, 26.3 **EP:** SE p. 578, Set B **ATS:** TE pp. 564B, 566B **MM:** FA/FF Level M
6.4.11 (D)	Select, use, and explain models to relate common fractions and mixed numbers (in halves, thirds, fourths, fifths, sixths, eighths and tenths); find equivalent fractions, mixed numbers, improper fractions, and decimals, and order fractions.	26.1, 26.3, 26.4	7	1/1	__/1	**SK:** 35	**R, P, PS:** 26.1, 26.3, 26.4 **EP:** SE p. 578, Sets A, C **ATS:** TE pp. 560B, 562, 566B, 568B, 570 **MM:** FA/NLM Levels M–O; FA/FF Levels L–N

Practice for the ISAT Test   xliv   **Individual Record Form**

Individual Record Form

Name _____   Date _____

TEST UNIT 8 (continued)

OBJECTIVES & LESSONS ASSESSED

Objective Number	Illinois Assessment Objectives Grade 4	SE/TE Lessons	Test Item(s)	Criterion Score	Student Score	Intervention	Prescription
6.4.12 (D)	Identify and generate equivalent forms of common decimals and fractions less than one whole (halves, quarters, fifths, and tenths).	26.3	8	1/1	__/1	**SK:** 35	**R, P, PS:** 26.3 **ATS:** TE p. 566B **MM:** FA/FF Level M
6.4.14 (B/C)	Add and subtract decimals through hundredths.	27.3, 27.4, 27.5	9, 10	2/2	__/2	**SK:** 38	**R, P, PS:** 27.3, 27.4, 27.5 **EP:** SE p. 598, Sets C–E **ATS:** TE pp. 588B, 590B, 592B, 594 **MM:** TNG/TT Level L; TNG/BB Levels F–I
6.4.15 (B/C)	Make estimates appropriate to a given situation with whole numbers, fractions, and decimals by knowing when to estimate, and select the appropriate type of estimate including overestimate, underestimate, and range of estimate, and select the appropriate method of estimation.	27.2	11, 12, 13	2/3	__/3	**PS:** 27	**R, P, PS:** 27.2 **EP:** SE p. 598, Set B **ATS:** TE p. 586B
6.4.29 (B/C)	Add and subtract with decimals expressed as tenths, using pictorial representations and monetary labels.	27.3, 27.4, 27.5	14	1/1	__/1	**PS:** 27	**R, P, PS:** 27.3, 27.4, 27.5 **EP:** SE p. 598, Sets C–E **ATS:** TE pp. 588B, 590B, 592B, 594 **MM:** TNG/TT Level L; TNG/BB Levels F–I

Individual Record Form — **Practice for the ISAT Test**

Individual Record Form

Name _____ Date _____

▶ **TEST UNIT 8** *(continued)*

OBJECTIVES & LESSONS ASSESSED

Goal 7: Measurement

Objective Number	Illinois Assessment Objectives Grade 4	SE/TE Lessons	Test Item(s)	Criterion Score	Student Score	Intervention	Prescription
7.4.01 (A/C) (OR)	Measure lengths to the nearest $\frac{1}{2}$ inch and $\frac{1}{2}$ cm with a ruler.	24.2	15	1/1	__/1	SK: 41	**R, P, PS:** 24.2 **EP:** SE p. 534, Set B **ATS:** TE pp. 522B, 524 **MM:** ISE/LL Level F
7.4.03 (A/C) (OR)	Convert both ways within systems without conversion charts (e.g., yards to feet, feet to inches, meters to centimeters, and hours to minutes).	24.3, 25.2, 25.6	16, 17	2/2	__/2	SK: 42	**R, P, PS:** 24.3, 25.2, 25.6 **EP:** SE pp. 534, Set C; 554, Sets B–C **ATS:** TE pp. 526B, 544B, 552B **MM:** TNG/TT Levels M, Q
7.4.04 (A/C)	Solve problems that require a knowledge of the following units: inches—down to $\frac{1}{2}, \frac{1}{4}$ feet and ?; feet, yards, miles, millimeters, centimeters, meters, and kilometers; weight/mass—ounces, pounds, tons, grams, and kilograms; liquid volume—cups, pints, quarts, gallons, milliliters, and liters; area—square units; temperature (Celsius and Fahrenheit units).	24.4, 24.5, 25.1, 25.3, 25.4	18	1/1	__/1	SK: 41–42	**R, P, PS:** 24.4, 24.5, 25.1, 25.3, 25.4 **EP:** SE p. 554, Set A **ATS:** TE pp. 528B, 530B, 540B, 542, 546B, 548B **MM:** TNG/TT Levels N–O, Q; ISE/LL Levels I–J
7.4.06 (B)	Choose the appropriate units (metric and U.S.) to estimate the length, liquid volume, and weight/mass of given objects.	24.1	19	1/1	__/1	SK: 41–42	**R, P, PS:** 24.1 **EP:** SE p. 534, Set A **ATS:** TE pp. 518B, 520 **MM:** ISE/LL Level E
7.4.07 (B)	Estimate the relative magnitudes of standard units (e.g., mm, cm, m).	24.6	20	1/1	__/1	PS: 24	**R, P, PS:** 24.6 **ATS:** TE p. 532B

Individual Record Form

Name _____ Date _____

TEST UNIT 8 (continued)

OBJECTIVES & LESSONS ASSESSED

Objective Number	Illinois Assessment Objectives Grade 4	SE/TE Lessons	Test Item(s)	Criterion Score	Student Score	Intervention	Prescription
7.4.08 (B)	Estimate standard measurements of length, weight, and capacity.	24.4, 24.5, 25.1, 25.3, 25.4	21	1/1	__/1	SK: 41–42	**R, P, PS:** 24.4, 24.5, 25.1, 25.3, 25.4 **EP:** SE p. 554, Set A **ATS:** TE pp. 528B, 530B, 540B, 542, 546B, 548B **MM:** TNG/TT Levels N–O, Q; ISE/LL Levels I–J
7.4.09 (A/C)	Perform simple unit conversions within a system of measurement (e.g., feet to inches, yards to feet).	24.3, 25.2	22, 23	2/2	__/2	SK: 41–42	**R, P, PS:** 24.3, 25.2 **EP:** SE pp. 534, Set C; 554, Set B **ATS:** TE pp. 526B, 544B **MM:** TNG/TT Levels M, Q
Extended-Response Items: Use rubric scoring							
6.4.14 (B/C)	Add and subtract decimals through hundredths.	27.3, 27.4, 27.5	24			SK: 38	**R, P, PS:** 27.3, 27.4, 27.5 **EP:** SE p. 598, Sets C–E **ATS:** TE pp. 588B, 590B, 592B, 594 **MM:** TNG/TT Level L; TNG/BB Levels F–I
7.4.04 (A/C)	Solve problems that require a knowledge of the following units: inches—down to $\frac{1}{2}, \frac{1}{4}$ feet and ?; feet, yards, miles, millimeters, centimeters, meters, and kilometers; weight/mass—ounces, pounds, tons, grams, and kilograms; liquid volume—cups, pints, quarts, gallons, milliliters, and liters; area—square units; temperature (Celsius and Fahrenheit units).	24.4, 24.5, 25.1, 25.3, 25.4	25			SK: 41–42	**R, P, PS:** 24.4, 24.5, 25.1, 25.3, 25.4 **EP:** SE p. 554, Set A **ATS:** TE pp. 528B, 530B, 540B, 542, 546B, 548B **MM:** TNG/TT Levels N–O, Q; ISE/LL Levels I–J

Individual Record Form   xlvii   **Practice for the ISAT Test**

Individual Record Form

Name _____ Date _____

TEST UNIT 9

OBJECTIVES & LESSONS ASSESSED

Objective Number	Illinois Assessment Objectives Grade 4	SE/TE Lessons	Test Item(s)	Criterion Score	Student Score	Intervention	Prescription
Goal 7: Measurement							
7.4.05 (A/C) (OR)	Calculate the area and perimeter of a rectangle, triangle, or irregular shape using diagrams, models, and grids or by measuring. Use the appropriate units in the response [e.g., square centimeter (cm^2), square meter (m^2), square inch (in^2), or square yard (yd^2)].	28.2, 29.2	1, 2, 3, 4, 5, 6, 15	5/7	__/7	SK: 43–44	**R, P, PS:** 28.2, 29.2 **EP:** SE pp. 622, Set A; 638, Set A **ATS:** TE pp. 614B, 616, 630B, 632 **MM:** ISE/PP Level R
Goal 9: Geometry							
9.4.01 (A)	Identify, describe and classify common three-dimensional geometric objects [e.g., cube and rectangular solids (prisms), sphere, pyramid, cone, and cylinder].	30.1	8, 13	2/2	__/2	SK: 53	**R, P, PS:** 30.1 **EP:** SE p. 656, Set A **ATS:** TE pp. 644B, 646 **MM:** ISE/FS Levels F–G
9.4.05 (B)	Identify common solid objects that are the components needed to make a more complex solid object.	30.2	9, 10, 11	2/3	__/3	SK: 53	**R, P, PS:** 30.2 **EP:** SE p. 656, Set B **ATS:** TE p. 648B **MM:** ISE/FS Levels H–I

Individual Record Form

Name _____ Date _____

▲ TEST UNIT 9 (continued)

OBJECTIVES & LESSONS ASSESSED

Objective Number	Illinois Assessment Objectives Grade 4	SE/TE Lessons	Test Item(s)	Criterion Score	Student Score	Intervention	Prescription
9.4.06 (B)	Identify prisms (including cubes) and pyramids in terms of the number and shape of faces, edges, and vertices.	30.1	7, 12, 14	2/3	___/3	**SK:** 53	**R, P, PS:** 30.1 **EP:** SE p. 656, Set A **ATS:** TE pp. 644B, 646 **MM:** ISE/FS Levels F–G
Extended-Response Items: Use rubric scoring							
7.4.05 (A/C) (OR)	Calculate the area and perimeter of a rectangle, triangle, or irregular shape using diagrams, models, and grids or by measuring. Use the appropriate units in the response [e.g., square centimeter (cm^2), square meter (m^2), square inch (in^2), or square yard (yd^2)].	28.2, 29.2	16, 17			**SK:** 43–44	**R, P, PS:** 28.2, 29.2 **EP:** SE pp. 622, Set A; 638, Set A **ATS:** TE pp. 614B, 616, 630B, 632 **MM:** ISE/PP Level R

VOCABULARY PRACTICE
Number Sense

Name _____

Write the word or words that best complete each sentence. Some words will not be used.

decimal	fraction	greater than	hundreds
hundredth	less than	millions	mixed number
period	place-value	tenth	thousands

1. The **decimal** equivalent for $\frac{1}{2}$ is 0.5.

2. The digit in the **millions** place in the number 12,345,678 is 2.

3. The three-digit group of 512 is in the thousands **period** in the number 6,512,904.

4. You can rewrite the whole number 5 as the **fraction** $\frac{15}{3}$.

5. The number 8.60 is **greater than** the number 8.5.

6. $4\frac{2}{3}$ is called a **mixed number**.

7. The number 615 is **less than** the number 621.

8. The digit in the **hundreds** place in the number 3,642,781 is 7.

9. When 8.15 is rounded to the nearest **tenth** it becomes 8.2.

10. The **place-value** position of the digit 6 in the number 162,195 is ten thousands.

Vocabulary 1 **Practice for the ISAT Test**

VOCABULARY PRACTICE
Number Sense

Name _____

Choose the term from the box at the right to make the statement true.

estimate
factor
inverse operation
multiple
product

1. Mel found the sum of two factors by multiplying them. **product**

2. John found a(n) exact answer for 32 times 18 by rounding the factors before he multiplied. **estimate**

3. Ann knew that 15 is a(n) factor of 3 because 3 times 5 equals 15. **multiple**

4. Elaine checked her division problem by using the Commutative Property, which is multiplication. **inverse operation**

Match each term to the correct definition. Record the numbers in the magic square. To check your work, make sure the sums for the columns, rows, and diagonals are all the same.

A	B	C
4	9	2
D	E	F
3	5	7
G	H	I
8	1	6

A. remainder
B. dividend
C. divisor
D. quotient
E. Associative Property of Multiplication
F. fact family
G. sum
H. divide
I. compatible numbers

1. to find how many items will be in each group, or to find how many groups
2. the number that divides the dividend
3. the answer, not including the remainder, that results from dividing
4. the amount left over when a number cannot be divided equally
5. when you group factors in different ways, the product is the same
6. numbers that are easy to compute mentally
7. related multiplication and division equations that use the same numbers
8. the answer when adding
9. the number that is divided

Practice for the ISAT Test 2 **Vocabulary**

VOCABULARY PRACTICE
Measurement

Name _____

Write the word or words that match each definition.

area	capacity	degrees Fahrenheit	elapsed time	gallon
kilogram	liter	perimeter	quart	volume

1. one of these equals 1,000 milliliters _liter_

2. number of squares that cover a flat surface _area_

3. amount of time that passes from the start of an activity to the end of that activity _elapsed time_

4. one of these equals two pints _quart_

5. a unit for measuring temperature _degrees Fahrenheit_

6. amount of space a solid figure takes up _volume_

7. one of these equals four quarts _gallon_

8. metric unit for measuring mass _kilogram_

9. distance around a figure _perimeter_

10. amount a container can hold _capacity_

Practice for the ISAT Test   4   Vocabulary

VOCABULARY PRACTICE
Measurement

Name _____

Circle the vocabulary term that describes all the other terms. Explain how the other terms in the list are related. _Possible answers are given._

1. decade
   minute
   (time)
   seconds

   Each is a unit of time.

2. ounce
   (weight)
   pound
   ton

   Each is a unit of weight.

3. inch
   yard
   (length)
   centimeter
   meter

   Each is a unit of length.

4. milliliter
   (capacity)
   cup
   gallon
   liter

   Each is a unit of capacity.

Answer each question for each measurement unit. Write _yes_ or _no_ in the chart. _Possible answers are given._

	Longer than your thumb?	Wider than your hand?	Longer than the distance from home to school?	Longer than your math book?
inch	no	no	no	no
foot	yes	yes	no	yes
yard	yes	yes	no	yes
mile	yes	yes	Answers may vary.	yes
millimeter	no	no	no	no
centimeter	no	no	no	no
meter	yes	yes	no	yes
kilometer	yes	yes	Answers may vary.	yes

Vocabulary   3   Practice for the ISAT Test

Practice for the ISAT Test   2   Vocabulary

VOCABULARY PRACTICE: Algebra

Name _____

Choose the correct word or words in parentheses to complete each sentence.

1. The first operation to perform in the expression $4 + 10 \div 2$ is __**division**__ (addition, division).

2. In $3 + n = 7$, n is a(n) __**variable**__ (operation, variable).

3. In the expression $9 \div 3 \times 2$, first find the __**quotient**__ (product, quotient).

4. Multiplication and division are __**inverse operations**__ (inverse operations, order of operations).

5. An example of an __**expression**__ (expression, equation) is $20 + n$.

6. $21 > 15 + x$ is an __**inequality**__ (inverse operation, inequality).

7. Marci receives an allowance of $2 per week. To find the total amount she will receive in 15 weeks, she should use the operation __**multiplication**__ (multiplication, division).

8. $15 - y = 12$ is an example of an __**equation**__ (expression, equation).

9. The correct __**order of operations**__ (order of operations, inverse operations) to use to find $12 - 6 \div 2$ is to divide and then subtract.

10. The symbols used to show which operations in an expression should be performed first are __**parentheses**__ (inequalities, parentheses).

Practice for the ISAT Test — 6 — Vocabulary

VOCABULARY PRACTICE: Algebra

Name _____

| equation | parentheses | equality |
| expression | variable | |

Write the term for each definition.

1. a mathematical sentence that uses the = sign.
   E **Q** U A **L** I T Y

2. a letter or symbol that stands for a number or numbers
   V A **R** I A B **L** E

3. a part of a number sentence that has numbers and operation signs but does not have an equal sign
   E **X** P R E **S** S I **O** N

4. a number sentence that shows that two quantities are equal
   E **Q** U **A** T I **O** N

Next, rearrange the letters from the boxes above to find the missing term in this sentence.

A **R** **U** **L** **E** for the input/output table is: Multiply by **8**.

x	$5 + 4$	$16 > 11$	$18 + 40 = 58$	n
$25 + 123 < 211$	$33 - a = 15$	h	$81 \div c$	$6 \times 7 = 42$
$22 \times y$	b	$g > 3{,}001$	$45 \div w = 5$	$60 - 15$

Sort the entries in the boxes below into the appropriate columns to tell whether each is a variable, an inequality, an expression, or an equation.

Variable	Inequality	Expression	Equation
x	$16 > 11$	$5 + 4$	$18 + 40 = 58$
n	$25 + 123 < 211$	$81 \div c$	$33 - a = 15$
h	$g > 3{,}001$	$22 \times y$	$6 \times 7 = 42$
b		$60 - 15$	$45 \div w = 5$

Vocabulary — 5 — Practice for the ISAT Test

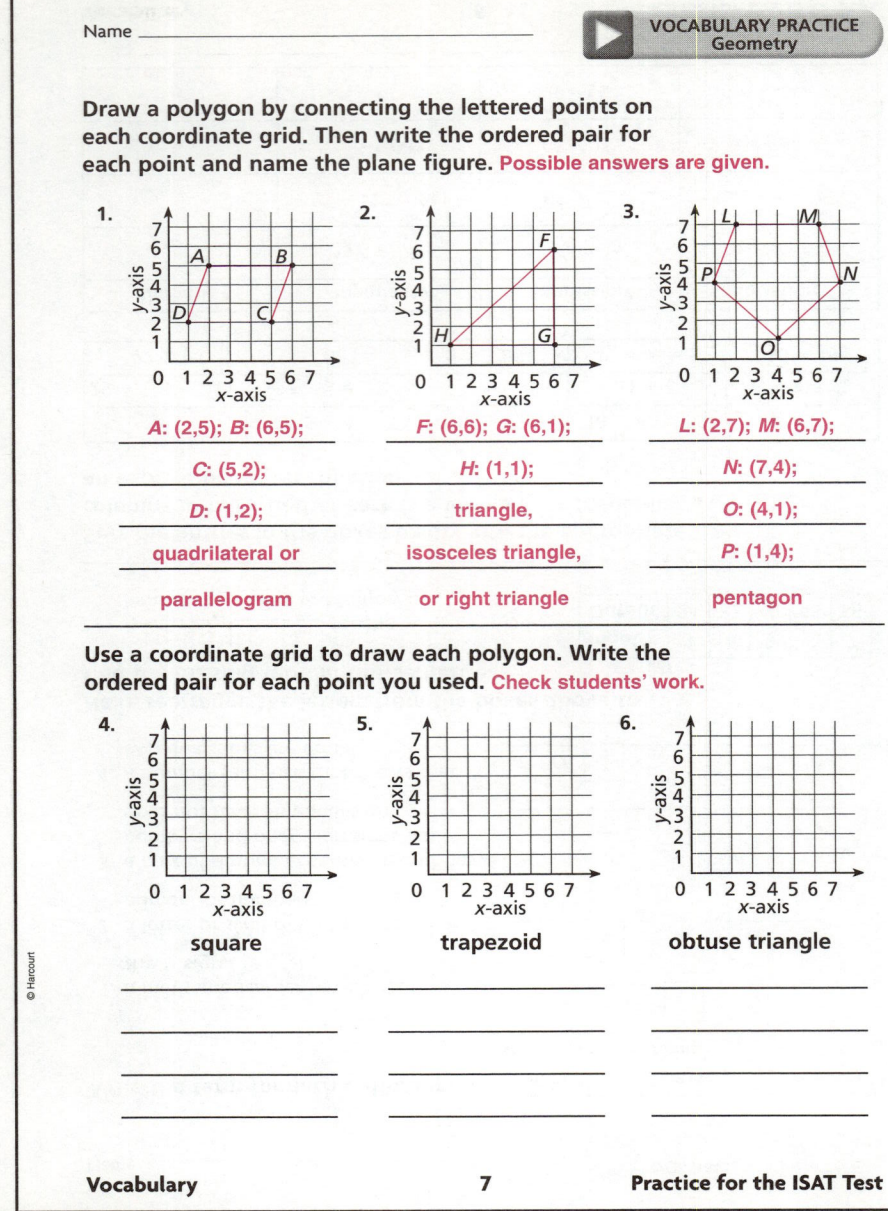

VOCABULARY PRACTICE
Geometry

Name _____

Match the clue to the shape.

1. I have 5 faces and 9 edges.

2. I have a curved surface, no edges, and no vertices.

3. I have 6 faces that are all squares.

4. I have 8 edges and 5 vertices.

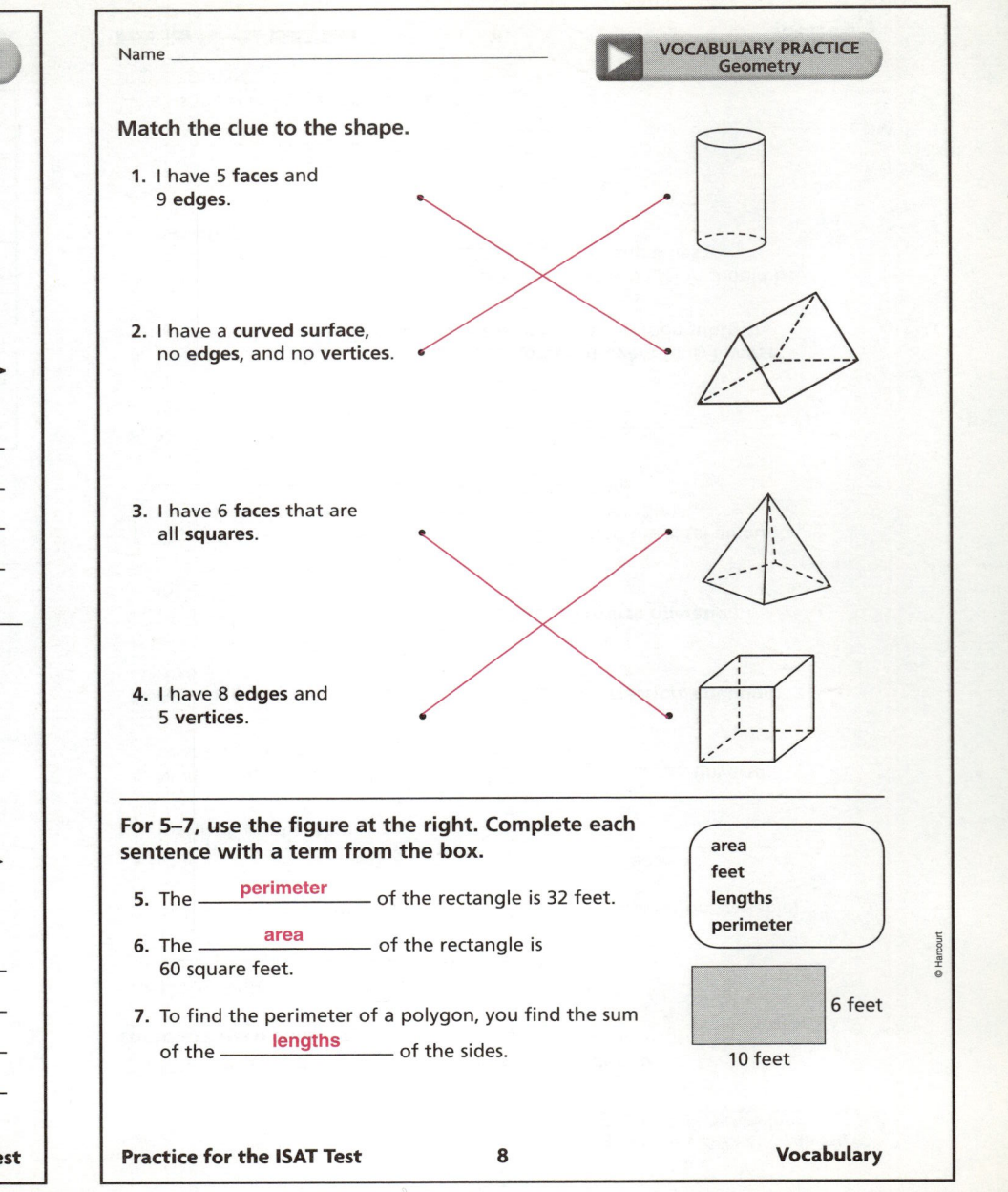

For 5–7, use the figure at the right. Complete each sentence with a term from the box.

| area |
| feet |
| lengths |
| perimeter |

6 feet / 10 feet

5. The **perimeter** of the rectangle is 32 feet.

6. The **area** of the rectangle is 60 square feet.

7. To find the perimeter of a polygon, you find the sum of the **lengths** of the sides.

Practice for the ISAT Test — 8 — Vocabulary

VOCABULARY PRACTICE
Geometry

Name _____

Draw a polygon by connecting the lettered points on each coordinate grid. Then write the ordered pair for each point and name the plane figure. Possible answers are given.

1.
A: (2,5); B: (6,5); C: (5,2); D: (1,2);
quadrilateral or parallelogram

2.
F: (6,6); G: (6,1); H: (1,1);
triangle, isosceles triangle, or right triangle

3.
L: (2,7); M: (6,7); N: (7,4); O: (4,1); P: (1,4);
pentagon

Use a coordinate grid to draw each polygon. Write the ordered pair for each point you used. Check students' work.

4. square

5. trapezoid

6. obtuse triangle

Vocabulary — 7 — Practice for the ISAT Test

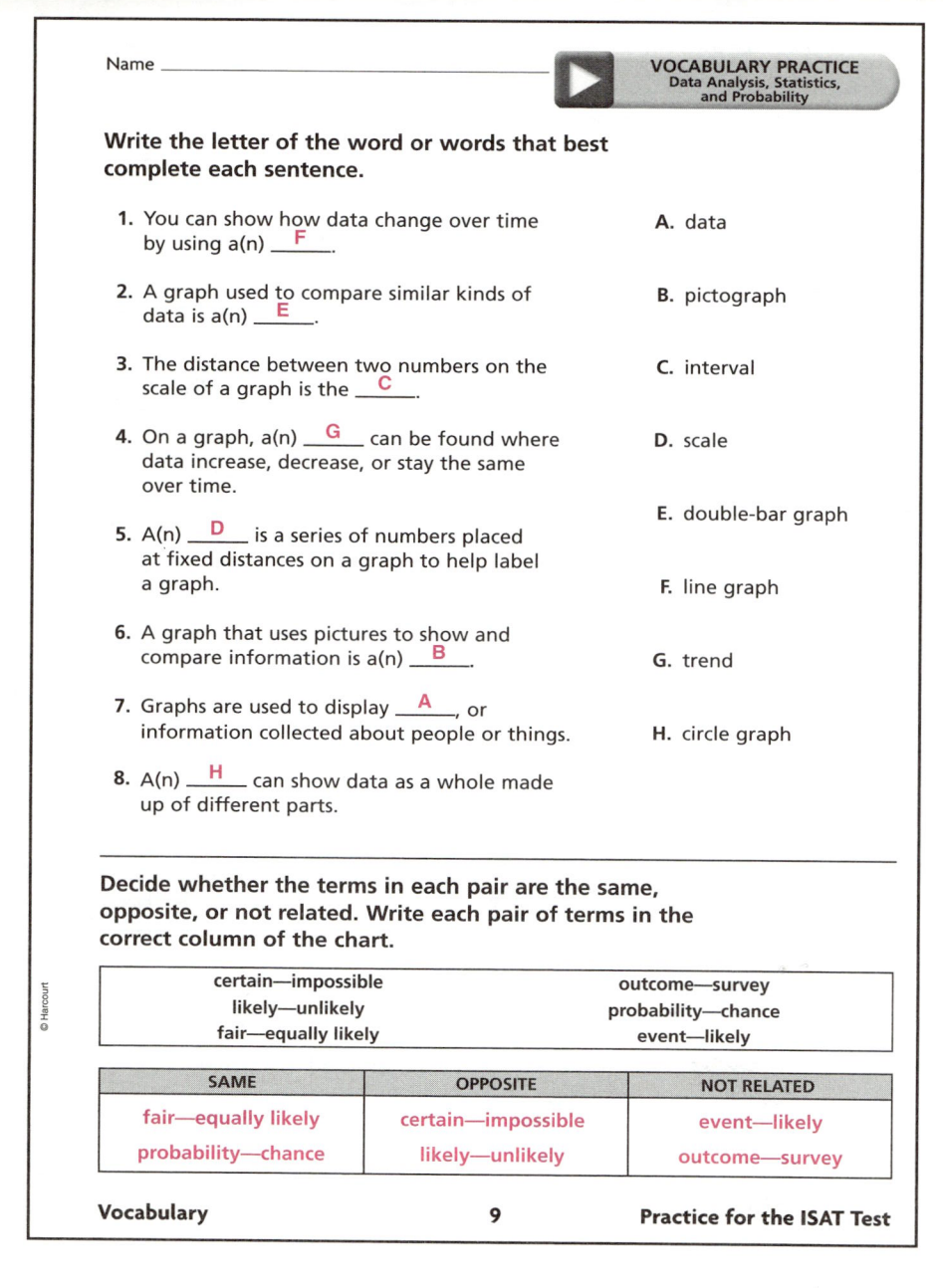

VOCABULARY PRACTICE
Data Analysis, Statistics, and Probability

Name _____

Write the letter of the word or words that best complete each sentence.

1. You can show how data change over time by using a(n) __F__.

2. A graph used to compare similar kinds of data is a(n) __E__.

3. The distance between two numbers on the scale of a graph is the __C__.

4. On a graph, a(n) __G__ can be found where data increase, decrease, or stay the same over time.

5. A(n) __D__ is a series of numbers placed at fixed distances on a graph to help label a graph.

6. A graph that uses pictures to show and compare information is a(n) __B__.

7. Graphs are used to display __A__, or information collected about people or things.

8. A(n) __H__ can show data as a whole made up of different parts.

A. data
B. pictograph
C. interval
D. scale
E. double-bar graph
F. line graph
G. trend
H. circle graph

Decide whether the terms in each pair are the same, opposite, or not related. Write each pair of terms in the correct column of the chart.

certain—impossible   outcome—survey
likely—unlikely   probability—chance
fair—equally likely   event—likely

SAME	OPPOSITE	NOT RELATED
fair—equally likely	certain—impossible	event—likely
probability—chance	likely—unlikely	outcome—survey

Vocabulary   9   Practice for the ISAT Test

VOCABULARY PRACTICE
Data Analysis, Statistics, and Probability

Name _____

Match the definitions to the correct terms. Record the numbers in the magic square. To check your work, find the sum of each row, column, and diagonal. Each of the sums should be the same.

A	B	C
2	7	6

D	E	F
9	5	1

G	H	I
4	3	8

1. cumulative frequency
2. interval
3. mean
4. median
5. mode(s)
6. outlier
7. range
8. scale
9. trends

A. the difference between two numbers on the scale of a graph

B. the difference between the greatest number and the least number in a set of data

C. a value separated from the rest of the data

D. on a graph, areas in which the data increase, decrease, or stay the same over time

E. the number(s) or item(s) that occur(s) most often in a set of data

F. a running total of items being counted

G. the middle number in an ordered set of data

H. the number found by dividing the sum of a set of numbers by the number of addends

I. a series of numbers placed at fixed distances on a graph to help label the graph

What is the sum for each row, column, and diagonal? __15__

Practice for the ISAT Test   10   Vocabulary

Name _____

1 If the sale price shown below is increased by one hundred dollars, which is the word form for the new price of the house?

HOUSE FOR SALE
$98,349

A ninety-eight thousand, three hundred forty-nine
B ninety-eight thousand, three hundred fifty-nine
C ninety-eight thousand, four hundred forty-nine
D ninety-nine thousand, three hundred forty-nine
E ninety-nine thousand, four hundred forty-nine

2 Which statement is true?

A 37,265 < 35,278
B 37,265 = 35,278
C 37,265 > 35,278
D 35,278 > 37,265
E 35,278 = 35,728

3 Carlos earns $195 each week. How much money does he earn each day if he works 5 days each week?

A $95   D $31
B $41   E $22
C $39

4 Look at the price of the car. What is the value of the digit 2?

$25,648

A 20   D 20,000
B 200   E 25,000
C 2,000

5 Jamal is reading a 423-page book. If he reads 11 pages per day, about how many days will it take him to finish the book?

A 4 days   D 40 days
B 10 days   E 400 days
C 20 days

6 What is the expanded form of 35,201?

A 30,000 + 5,000 + 20 + 1
B 30,000 + 5,000 + 200 + 1
C 30,000 + 500 + 20 + 1
D 30,000 + 500 + 201
E 3,000 + 500 + 20 + 1

7 Martin's Pet Store has 58 birds for sale. On Saturday, 7 customers bought 2 birds each. How many birds are left?

A 14 birds   D 102 birds
B 44 birds   E 132 birds
C 49 birds

Name _____

8 Karen is making a division model to show the inverse of the multiplication model shown below.

If she makes 6 groups, how many counters will be in each group?

A 5 counters   D 8 counters
B 6 counters   E 9 counters
C 7 counters

9 Gina used place-value materials to model a division problem. Which division problem does her model represent?

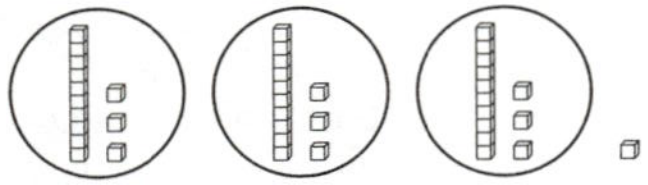

A 40 ÷ 4 = 10
B 40 ÷ 3 = 13 r1
C 39 ÷ 3 = 13
D 39 ÷ 3 = 13 r1
E 39 ÷ 3 = 10 r3

10 A bookstore receives 787 books in 10 cartons. There are about the same number of books in each carton. About how many books are in each carton?

A about 50 books
B about 60 books
C about 70 books
D about 80 books
E about 90 books

11 Lyle is walking home from the library. He has walked $\frac{1}{4}$ mile. He lives $\frac{7}{8}$ mile from the library.

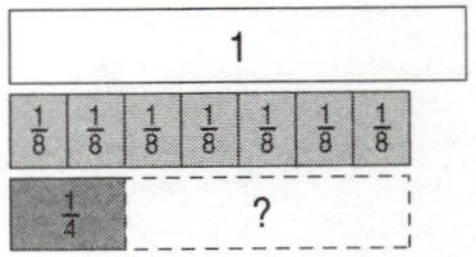

Use the fraction bars to find out how far Lyle still has to walk to get home.

A $\frac{1}{8}$ mile   D $\frac{5}{8}$ mile
B $\frac{1}{4}$ mile   E $\frac{3}{4}$ mile
C $\frac{3}{8}$ mile

12 The number of dozens of eggs is a mixed number. Which mixed number is greater than the number of dozens of eggs shown?

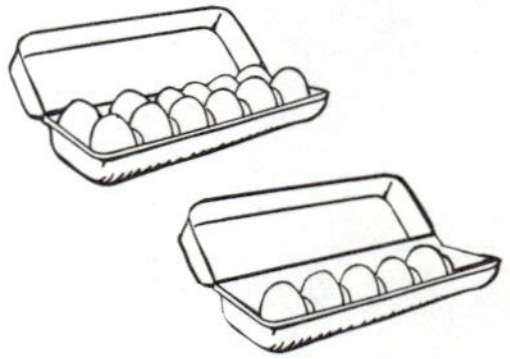

A $1\frac{1}{12}$ dozen
B $1\frac{2}{12}$ dozen
C $1\frac{3}{12}$ dozen
D $1\frac{5}{12}$ dozen
E $1\frac{7}{12}$ dozen

GO ON TO THE NEXT PAGE.

Name _____

13 Which number completes the pattern shown?

480 ÷ 6 = 80
4,800 ÷ 6 = 800
48,000 ÷ 6 = ■

A 8
B 80
C 800
D 8,000
E 80,000

14 Cassandra is painting on canvas. She paints 5/8 of the canvas in a striped design, and 1/4 of the canvas solid orange. Which fraction of Cassandra's painting is either striped or orange?

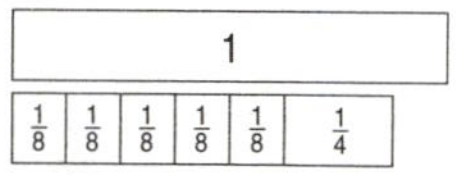

A 3/8
B 5/12
C 1/2
D 7/12
E 7/8

15 The large rectangle represents 1. Which fraction is equivalent to the shaded part of the rectangle?

A 1/4
B 1/2
C 3/4
D 5/16
E 11/16

16 Blake's house is seventy-eight hundredths mile from school. Write the number of miles from Blake's house to his school in standard form.

A 780
B 78
C 7.8
D 0.78
E 0.078

17 Sallie earned an A on 3 out of 10 math quizzes. Which decimal shows how often she earned an A?

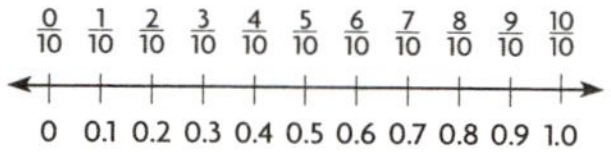

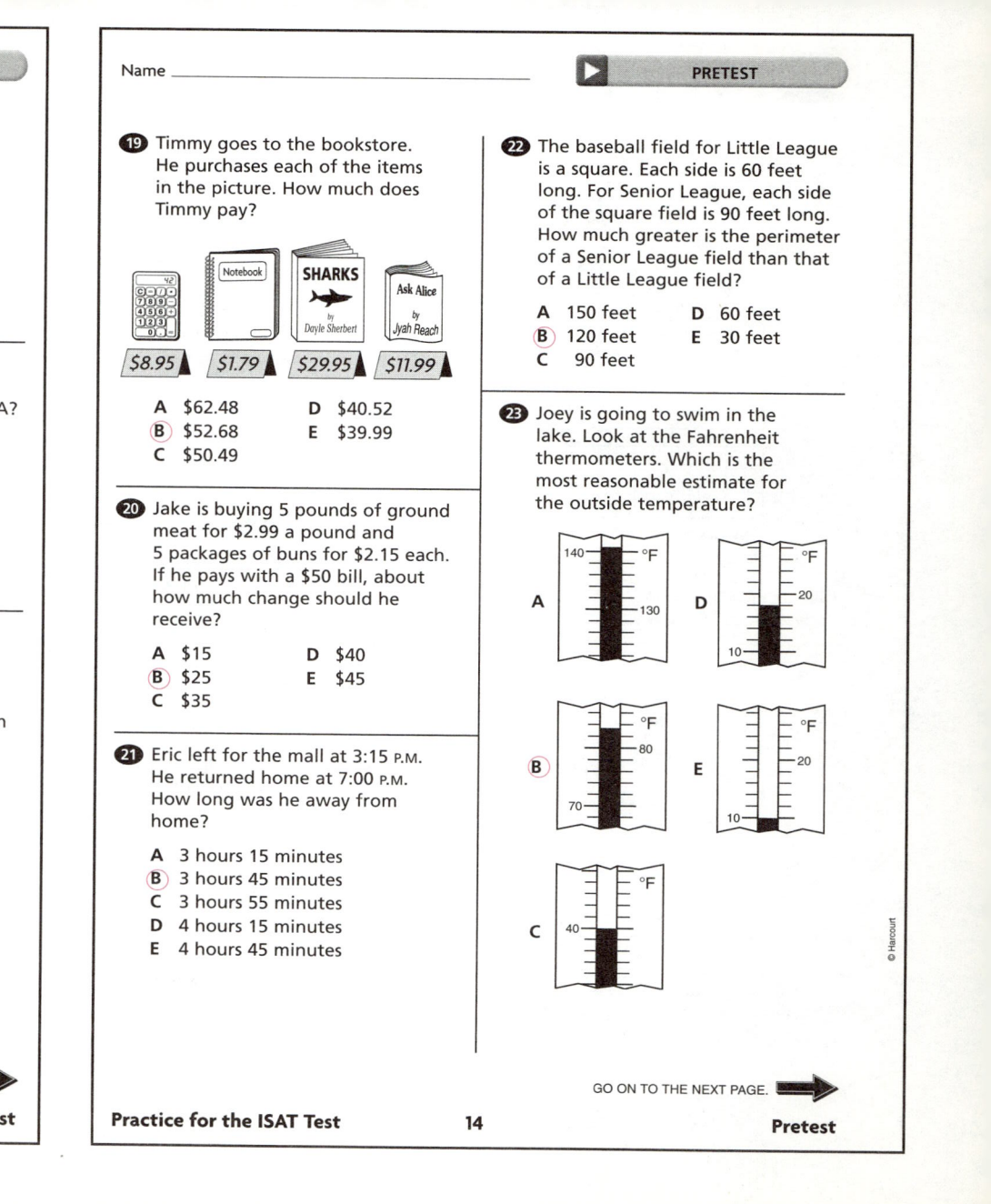

A 7.0
B 3.10
C 3.0
D 0.7
E 0.3

18 The table shows the amount of money each employee at Candy's Pet Store earns per hour. How much more money does the highest-paid employee make than the lowest-paid employee?

EARNINGS FOR CANDY'S EMPLOYEES	
Employee	Pay Per Hour
Mandy	$7.65
Sandy	$11.32
Andy	$6.79
Randy	$10.28
Brandy	$9.26

A $18.11
B $5.53
C $4.53
D $4.47
E $3.49

Pretest   13   Practice for the ISAT Test

Name _____

19 Timmy goes to the bookstore. He purchases each of the items in the picture. How much does Timmy pay?

$8.95 $1.79 $29.95 $11.99

A $62.48
B $52.68
C $50.49
D $40.52
E $39.99

20 Jake is buying 5 pounds of ground meat for $2.99 a pound and 5 packages of buns for $2.15 each. If he pays with a $50 bill, about how much change should he receive?

A $15
B $25
C $35
D $40
E $45

21 Eric left for the mall at 3:15 P.M. He returned home at 7:00 P.M. How long was he away from home?

A 3 hours 15 minutes
B 3 hours 45 minutes
C 3 hours 55 minutes
D 4 hours 15 minutes
E 4 hours 45 minutes

22 The baseball field for Little League is a square. Each side is 60 feet long. For Senior League, each side of the square field is 90 feet long. How much greater is the perimeter of a Senior League field than that of a Little League field?

A 150 feet
B 120 feet
C 90 feet
D 60 feet
E 30 feet

23 Joey is going to swim in the lake. Look at the Fahrenheit thermometers. Which is the most reasonable estimate for the outside temperature?

A   **B**   C   D   E

Practice for the ISAT Test   14   Pretest

Name _____

24 How many more points did Michelle score than Toby?

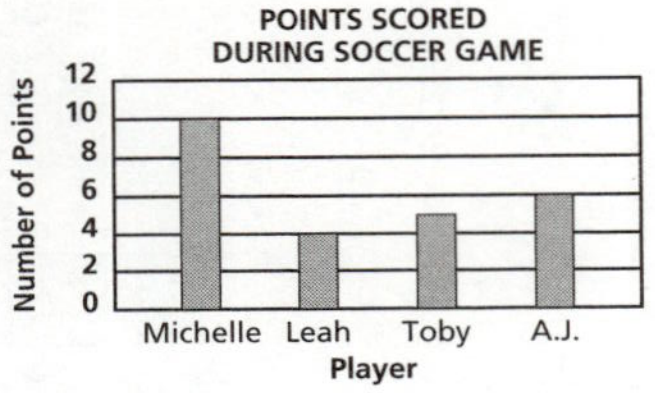

POINTS SCORED DURING SOCCER GAME

A 10 points   D 4 points
B 6 points   E 2 points
C 5 points ⊙

25 One packet of chicken broth has a mass of 10 grams. How many packets of chicken broth would have a mass of 1 kilogram?

A 1 packet   D 1,000 packets
B 10 packets   E 10,000 packets
C 100 packets ⊙

26 Walter grew a giant pumpkin that won a blue ribbon at the county fair. If the mass of the small pumpkin is 12 kilograms, about how much is the mass of Walter's pumpkin?

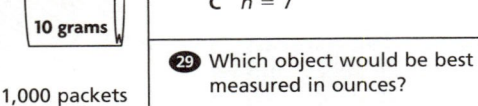

A 60 grams
B 600 grams
C 6 kilograms
D 60 kilograms ⊙
E 600 kilograms

27 This table shows how many quarts are in a gallon. Which is the missing number in the table?

Quarts	4	8	12	16	20
Gallon(s)	1	2	3	4	5

A 22 quarts   D 28 quarts
B 24 quarts ⊙   E 30 quarts
C 25 quarts

28 Use mental math. Which value for the variable makes the equation true?

$25 - h = 19$

A $h = 45$ D $h = 6$ ⊙
B $h = 44$ E $h = 4$
C $h = 7$

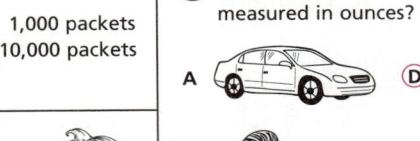

29 Which object would be best measured in ounces?

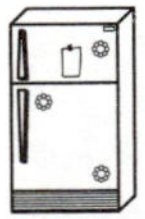

A (car)
B (tiger)
C (refrigerator)
D (baseball) ⊙
E (bicycle)

30 Mitch wants to carpet a living room, dining room, and hall with the same wall-to-wall carpeting.

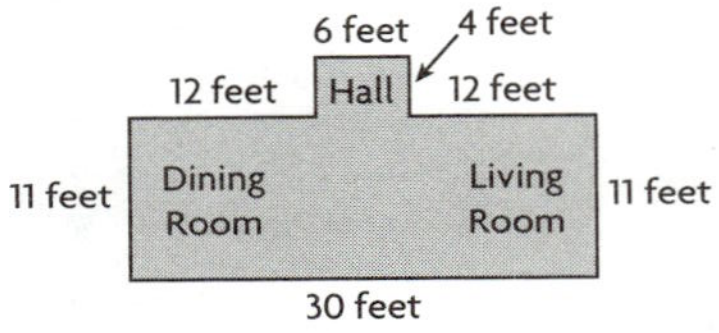

Use the diagram. How much carpeting will he need?

A 90 square feet
B 132 square feet
C 330 square feet
D 354 square feet ⊙
E 360 square feet

31 Find a rule for the pattern in the input/output table.

Input c	Output d
5	10
8	13
15	20
20	25

What is the rule written as an equation?

A $d + 5 = c$
B $c + 5 = d$ ⊙
C $c - 5 = d$
D $c + d = 5$
E $2 \times c = d$

32 Tari is going to spin the spinner. How likely is it that the pointer will stop on a number greater than 1?

A certain
B likely ⊙
C equally likely
D unlikely
E impossible

33 Based on the pattern in the graph, how much money will William earn if he works 10 hours?

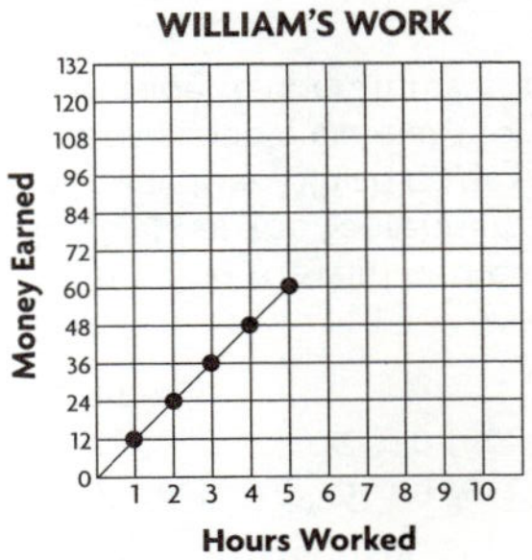

WILLIAM'S WORK

A $84
B $96
C $108
D $120 ⊙
E $132

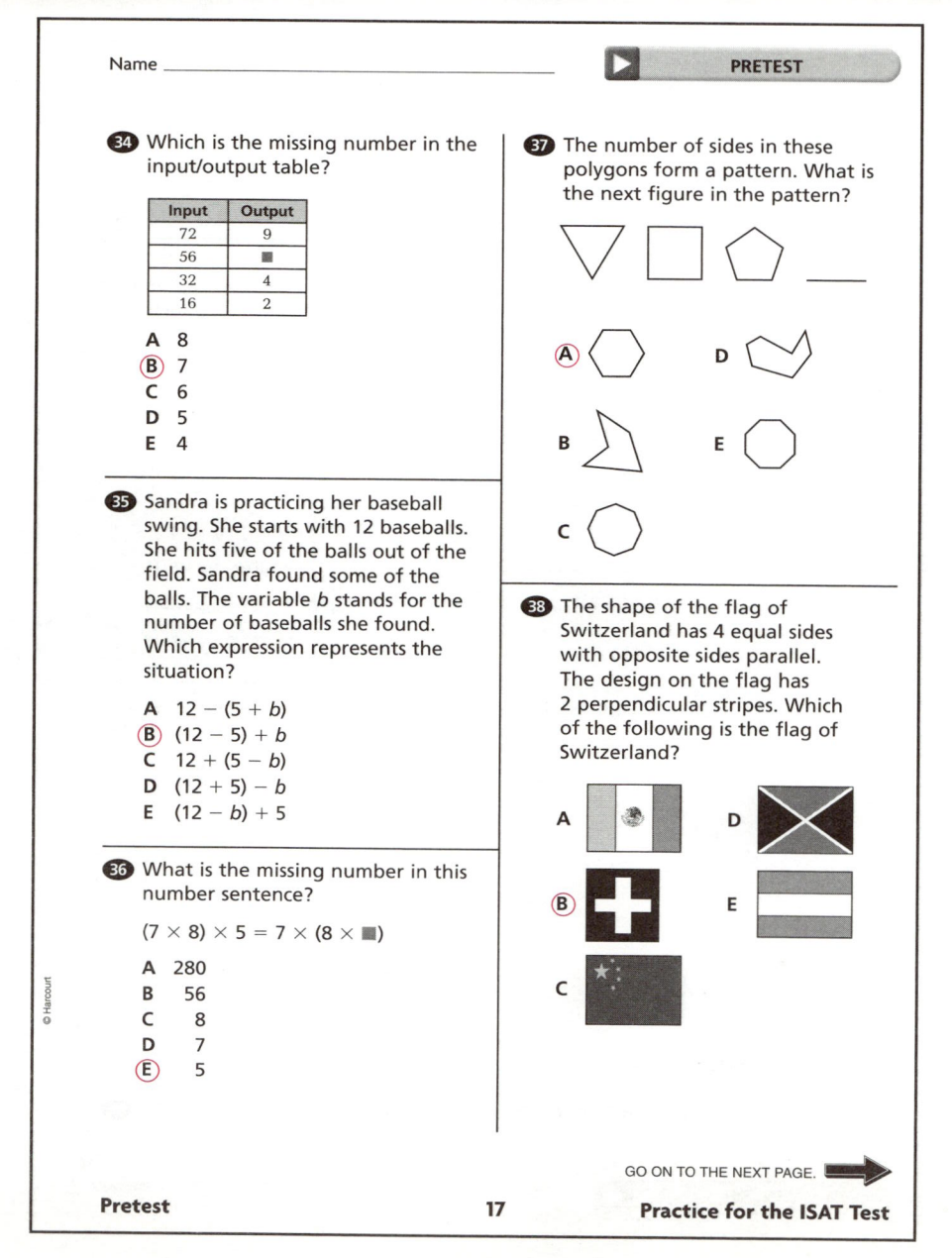

Name _____

34 Which is the missing number in the input/output table?

Input	Output
72	9
56	■
32	4
16	2

A 8
(B) 7
C 6
D 5
E 4

35 Sandra is practicing her baseball swing. She starts with 12 baseballs. She hits five of the balls out of the field. Sandra found some of the balls. The variable *b* stands for the number of baseballs she found. Which expression represents the situation?

A $12 - (5 + b)$
(B) $(12 - 5) + b$
C $12 + (5 - b)$
D $(12 + 5) - b$
E $(12 - b) + 5$

36 What is the missing number in this number sentence?

$(7 \times 8) \times 5 = 7 \times (8 \times \blacksquare)$

A 280
B 56
C 8
D 7
(E) 5

37 The number of sides in these polygons form a pattern. What is the next figure in the pattern?

(A)   B   C

38 The shape of the flag of Switzerland has 4 equal sides with opposite sides parallel. The design on the flag has 2 perpendicular stripes. Which of the following is the flag of Switzerland?

A (B) C D E

GO ON TO THE NEXT PAGE.

Name _____

39 Lisa walks 4 blocks west. Then she walks 5 blocks south, 3 blocks east, 2 blocks north, and 7 blocks west. How many blocks does Lisa walk before she crosses her own path?

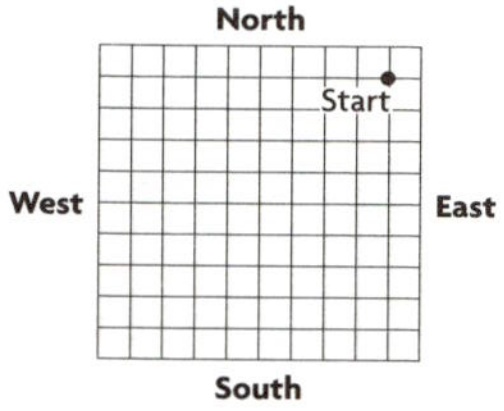

A 14 blocks
B 16 blocks
(C) 17 blocks
D 20 blocks
E 21 blocks

40 Roseanne is visiting Washington School for the first time. Where is the school located?

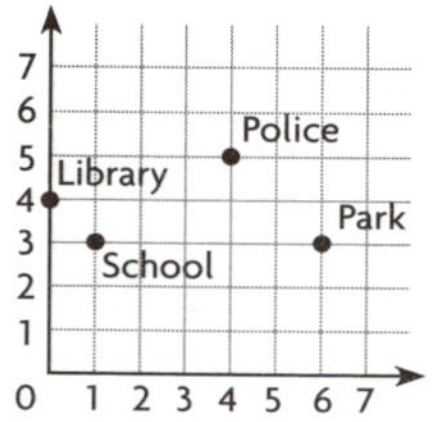

(A) (1,3)
B (3,1)
C (0,4)
D (4,0)
E (3,2)

41 Bob's dog, Honey, is now 6 months old. This graph shows how much Honey weighed each month. Which question can be answered by reading the graph?

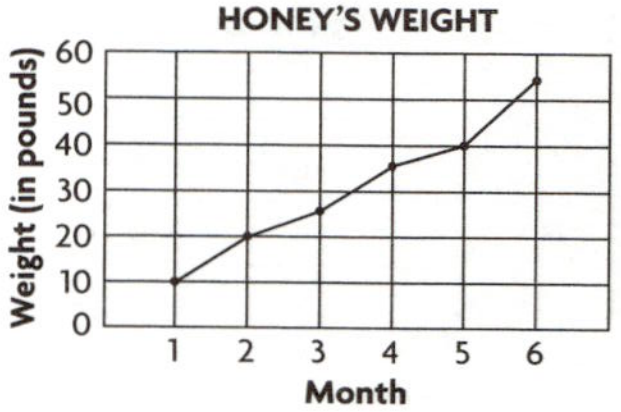

HONEY'S WEIGHT

A What kind of dog is Honey?
(B) How much weight has Honey gained since she was 1 month old?
C How much did Honey weigh when she was born?
D How tall is Honey?
E Honey will weigh 65 pounds when she is 7 months old.

42 This table shows how many pennies Shawn saved last week. Which is the median number of pennies saved?

Number of Pennies Saved

Sun	Mon	Tue	Wed	Thu	Fri	Sat
12	15	18	12	13	11	16

A 11   D 14
B 12   E 15
(C) 13

GO ON TO THE NEXT PAGE.

Name _____

43 At the China Café, Ling ordered egg drop soup, vegetable lo mein, chicken wings, and juice. How much was her bill?

China Café
Egg-drop Soup	$2.35
Pork Stir-fry	$6.99
Beef Stir-fry	$7.49
Vegetable Lo Mein	$5.75
Chicken Wings	$4.79
Juice	$1.25
Tea	$0.95

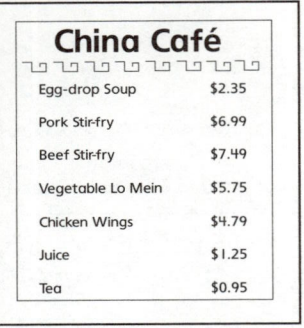

Estimate the answer by rounding the cost of each item to the nearest dollar.

Answer: $ __14__

Now, find the actual cost.
Show all work.

2.35 + 5.75 + 4.79 + 1.25 = 14.14

Answer: $ __14.14__

Is your answer reasonable? On the lines below, explain how you know.
Possible explanation: Yes. I rounded the price of each item ordered and then added to get an estimate. My answer is very close to my estimate of $14, so I know that my answer is reasonable.

GO ON TO THE NEXT PAGE.

Pretest   19   Practice for the ISAT Test

Name _____

44 Matt wants to plant a rectangular garden with an area of 30 square feet. A side of each unit square in the grid stands for 1 foot.

Draw three different rectangles, each of which has an area of 30 square feet. Possible rectangles are 1 foot by 30 feet, 2 feet by 15 feet, 3 feet by 10 feet, and 5 feet by 6 feet.

Make a table to show the width, length, perimeter, and area of each rectangle that you drew. Possible rectangles are given.

Rectangle	Width (in feet)	Length (in feet)	Perimeter (in feet)	Area (in square feet)
1	1	30	62	30
2	2	15	34	30
3	3	10	26	30
4	5	6	22	30

For which rectangle would Matt need to buy the least amount of fencing to enclose the garden? Explain how you know.
Possible answer: He would need the least amount of fencing for the garden that has the smallest perimeter.

STOP

Practice for the ISAT Test   20   Pretest

Name _____

TEST • UNIT 1

① Which list is ordered from greatest to least?

A 127,552; 152,875; 125,758
Ⓑ 152,875; 127,552; 125,758
C 152,875; 125,758; 127,552
D 125,758; 127,552; 152,875
E 125,758; 152,875; 127,552

② The table shows the populations of the five least populated states. Which state is the least populated?

LEAST POPULATED STATES	
State	Population
Alaska	626,932
Connecticut	560,381
North Dakota	642,200
Vermont	608,827
Wyoming	493,782

A Alaska
B Connecticut
C North Dakota
D Vermont
Ⓔ Wyoming

③ If 75,253 is increased by ten thousand, what is the new number?

Ⓐ 85,253
B 76,253
C 75,353
D 75,263
E 75,252

④ Tommy bought a car that was reduced in price by $10,000. How much did he pay for the car?

Price $42,385

A fifty-two thousand, three hundred eighty-five dollars
B fifty-one thousand, three hundred eighty-five dollars
C forty-one thousand, three hundred eighty-five dollars
D thirty-two thousand, two hundred eighty-five dollars
Ⓔ thirty-two thousand, three hundred eighty-five dollars

⑤ Which statement is true?

Ⓐ 58,329 > 58,239
B 58,329 = 58,239
C 58,329 < 58,239
D 58,239 > 58,329
E 58,239 = 58,329

⑥ Fans at a baseball game bought 1,186 hot dogs on Saturday. On Sunday, the fans bought 2,063 hot dogs. How many hot dogs were bought on Saturday and Sunday?

Ⓐ 3,249 hot dogs
B 3,219 hot dogs
C 3,149 hot dogs
D 3,119 hot dogs
E 877 hot dogs

GO ON TO THE NEXT PAGE.

Test • Unit 1 21 Practice for the ISAT Test

---

⑦ Pops Sports sold 42 footballs and 27 basketballs last week. About how many balls were sold in all?

A about 100 balls
B about 90 balls
C about 80 balls
Ⓓ about 70 balls
E about 60 balls

⑧ About how many acres larger is Natural Bridges than Walnut Canyon?

AREAS OF U.S. MONUMENTS	
Monument	Area (in acres)
Fossil Butte	8,198
Jewel Cave	1,274
Natural Bridges	7,636
Walnut Canyon	3,579

A about 10,000 acres
B about 5,000 acres
Ⓒ about 4,000 acres
D about 3,000 acres
E about 2,000 acres

⑨ Mr. Marshall is a pilot. Last month he flew a total distance of about 8,900 miles. This number was rounded to the nearest hundred miles. How many miles could he have flown last month?

A 7,829 miles   Ⓓ 8,874 miles
B 7,999 miles   E 8,969 miles
C 8,829 miles

⑩ Josh wants to buy a fishing boat and a trailer. He figures that the total cost of the boat and the trailer will be $9,595. Which estimate would help you decide whether his calculations are reasonable?

Boat—$6,205
Trailer—$1,995

A $9,200; His answer is reasonable.
B $9,000; His answer is reasonable.
C $8,500; His answer is not reasonable.
Ⓓ $8,200; His answer is not reasonable.
E $7,900; His answer is not reasonable.

⑪ What is the standard form for the number two thousand, ninety-three?

A 29,300   D 293
B 2,930   E 239
Ⓒ 2,093

GO ON TO THE NEXT PAGE.

Test • Unit 1 22 Practice for the ISAT Test

Name _____

TEST • UNIT 1

12 Mr. Taskin is selling all his merchandise and closing his store. His merchandise has a value of $87,892. He sold $31,358 worth of merchandise. About how much is the value of the merchandise he has left?

A $30,000
B $40,000
(C) $60,000
D $80,000
E $120,000

13 Jenny buys shampoo for $6 and a curling iron for $29. She writes the number sentence below to find out how much more she pays for the curling iron. Which term best describes Jenny's number sentence?

$29 − $6 = ■

A sum
B product
(C) difference
D quotient
E estimate

14 The table below shows the number of passengers on the morning train from Monday through Wednesday.

TRAIN PASSENGERS	
Day	Passengers
Monday	342
Tuesday	324
Wednesday	350

Which set of numbers is ordered from greatest to least?

A 342 < 324 < 350
B 324 > 342 > 350
(C) 350 > 342 > 324
D 350 < 342 < 324
E 342 < 350 < 324

15 The odometer on Don's car shows 57,325 miles. What is the expanded form for the number 57,325?

(A) 50,000 + 7,000 + 300 + 20 + 5
B 57,000 + 300 + 205
C 50,000 + 7,000 + 30 + 20 + 5
D 50,000 + 700 + 30 + 25
E 57,000 + 7,000 + 300 + 20 + 5

16 Which number is not equal to the others?

A 20,375
B 20,000 + 300 + 70 + 5
(C) twenty-three thousand, seventy-five
D 19,375 + 1,000
E 20,000 + 375

17 Use the number line. Which number is less than 6,720?

(number line from 6,700 to 6,750 with marks at 6,713, 6,723, 6,730, 6,734, 6,740, 6,745, 6,750)

(A) 6,713
B 6,723
C 6,730
D 6,734
E 6,745

18 Use mental math. Which value for the variable x makes the equation true?

$x + 7 = 21$

A $x = 28$
B $x = 27$
(C) $x = 14$
D $x = 13$
E $x = 3$

GO ON TO THE NEXT PAGE.

Name _____

19 Joe has collected 28 of the 50 state quarters. How many more quarters does he need to collect to have all 50 quarters?

$q + 28 = 50$

A $q = 12$
(B) $q = 22$
C $q = 32$
D $q = 38$
E $q = 42$

20 Chris and Andre were playing tennis. They started with 9 tennis balls. The variable b stands for the number of tennis balls that went over the fence. Chris and Andre have 6 tennis balls left. Which equation represents the situation?

(A) $9 − b = 6$
B $b − 9 = 6$
C $6 − b = 9$
D $9 + 6 = b$
E $b − 6 = 9$

21 Christina had 12 DVDs. She received more DVDs for her birthday. Now she has 18 DVDs. The variable d stands for the number of DVDs Christina received for her birthday. Which equation best describes the situation?

A $d − 12 = 18$
(B) $12 + d = 18$
C $18 + d = 12$
D $12 − d = 18$
E $d − 18 = 12$

TEST • UNIT 1

22 A flower shop charges the same delivery fee for each arrangement. The total price for a $17 arrangement is $25, and the total price for a $25 arrangement is $33. What is the total price for a $41 arrangement? Use the input/output table.

Input	Output
17	25
25	33
33	41
41	■

A $33
B $41
C $45
(D) $49
E $50

23 Find a rule for the pattern in the input/output table.

Input x	Output y
22	14
27	19
34	26
41	33

What is the rule written as an equation?

A $x + 8 = y$
(B) $x − 8 = y$
C $y − 8 = x$
D $y = 8 − x$
E $x + y = 8$

24 Which number is next in the pattern?

17, 26, 35, 44, ■

A 64
B 62
C 55
(D) 53
E 51

GO ON TO THE NEXT PAGE.

Name _____

25 The five tallest mountains in the United States, Canada, and Mexico in order of height from the tallest to the shortest are McKinley, Logan, Pico de Orizaba, St. Elias, and Popocatepetl. The heights of the mountains in random order are 19,551 feet, 18,008 feet, 20,320 feet, 18,555 feet, and 17,930 feet. Make a table to show the mountains and their heights in order from tallest to shortest.

Tallest Mountains in U.S., Canada, Mexico

Mountain	Height (in feet)
McKinley	20,320
Logan	19,551
Pico de Orizaba	18,555
St. Elias	18,008
Popocatepetl	17,930

On the lines below, explain using place values how someone would know that Logan is taller than Pico de Orizaba.

Possible explanation: The person would compare the heights of the two mountains. The digit 9 in the thousands place of 19,551 is greater than the digit 8 in the thousands place of 18,555. So, 19,551 > 18,555.

Test • Unit 1   25   Practice for the ISAT Test

GO ON TO THE NEXT PAGE.

Name _____

26 Gina has 15 pennies in her bank. She puts in 5 more pennies and then takes out 7 pennies. Her brother Jeff has 6 pennies in his bank. How many more pennies does he need to have the same number of pennies that Gina has? Explain what steps you could use to solve the problem. Then find the answer.

Possible answer: Start with 15 pennies. Add 5 more pennies and then take away 7 pennies. Take away 6 more pennies. Count the number of pennies left. That is how many pennies Jeff needs.

_____
_____

Now find the answer.

Answer: Jeff needs ___7___ more pennies.

STOP

Practice for the ISAT Test   26   Test • Unit 1

Test • Unit 1   13   Practice for the ISAT Test

Name _____

TEST • UNIT 2

1. Brad volunteered to plant new flowers at the park. The table shows the time he started working and how long he worked. At what time did Brad finish working at the park?

Start Time	End Time	Elapsed Time
8:45 A.M.		4 hr 17 min

- A 12:02 P.M.
- B 12:12 P.M.
- C 12:45 P.M.
- (D) 1:02 P.M.
- E 1:12 P.M.

2. Lauren leaves for work at 8:14 A.M. She returns home at 5:57 P.M. How long is she away from home?

- A 2 hours 17 minutes
- B 3 hours 53 minutes
- C 8 hours 43 minutes
- (D) 9 hours 43 minutes
- E 10 hours 43 minutes

3. Ms. Roberts arrived by train in Oklahoma City at 6:45 P.M. She had been on the train for 8 hours 55 minutes. At what time did Ms. Roberts board the train?

- A 3:40 A.M.
- B 3:45 A.M.
- (C) 9:50 A.M.
- D 9:55 A.M.
- E 10:50 A.M.

4. The table shows the maximum life spans of four animals. Suppose you wanted to make a graph to represent the data. Which type of graph would you use?

LIFE SPANS (IN YEARS)

Squirrel	Cat	Dog	Moose
23	28	20	27

- A line graph
- B line plot
- C double-bar graph
- (D) pictograph
- E double-line graph

5. This table shows the average cost of taking a pet to a veterinarian. Which conclusion cannot be made about the data?

AVERAGE COST AT VET (bar graph: Dog, Cat, Bird, Rabbit)

- A The cost is greater for a dog than for a cat.
- B The cost is greater for a bird than for a dog.
- (C) The cost for a bird is twice as much as the cost for a rabbit.
- D The cost for a bird is the greatest.
- E The cost for a cat is less than the cost for a rabbit.

GO ON TO THE NEXT PAGE.

Name _____

TEST • UNIT 2

6. Which kind of graph would be best to keep a record of monthly rainfall in Chicago?

- (A) line graph
- B circle graph
- C pictograph
- D line plot
- E double-bar graph

7. Which kind of graph would be best to use to compare the number of minutes Jenny read each night last week?

MINUTES JENNY READ
25, 45, 40, 50, 30, 35, 25

- A line graph
- B circle graph
- C double-bar graph
- (D) bar graph
- E line plot

8. Dana kept a tally of the scissor-tailed flycatchers in her yard. Then she made a frequency table. How many flycatchers did she see on Monday and Tuesday?

SCISSOR-TAILED FLYCATCHERS SEEN

Day	Frequency (Number of Birds)	Cumulative Frequency
Monday	14	14
Tuesday	8	
Wednesday	12	34
Thursday	9	43

- A 8 flycatchers
- B 14 flycatchers
- (C) 22 flycatchers
- D 36 flycatchers
- E 43 flycatchers

9. Which conclusion cannot be made about the data?

BOB'S WEIGHT LIFTING (line graph, Weeks 1–5, Weight in pounds 0–80)

- A Bob has increased the weight each week.
- B Bob lifted more weight in Week 4 than in Week 3.
- (C) Bob will decrease the weight in Week 6.
- D Bob increased the weight by 10 pounds each week.
- E Bob lifted twice as much weight in Week 3 as in Week 1.

10. A park ranger measured the height of a stream every 15 minutes during a heavy rainstorm. He recorded the data in the table below.

Time (minutes)	Height (inches)
15	2
30	4
45	6
60	8

Based on the data in the table, how fast was the water level rising?

- A 1 inch every 5 minutes
- B 1 inch every 15 minutes
- C 2 inches every 10 minutes
- (D) 2 inches every 15 minutes
- E 2 inches every 30 minutes

GO ON TO THE NEXT PAGE.

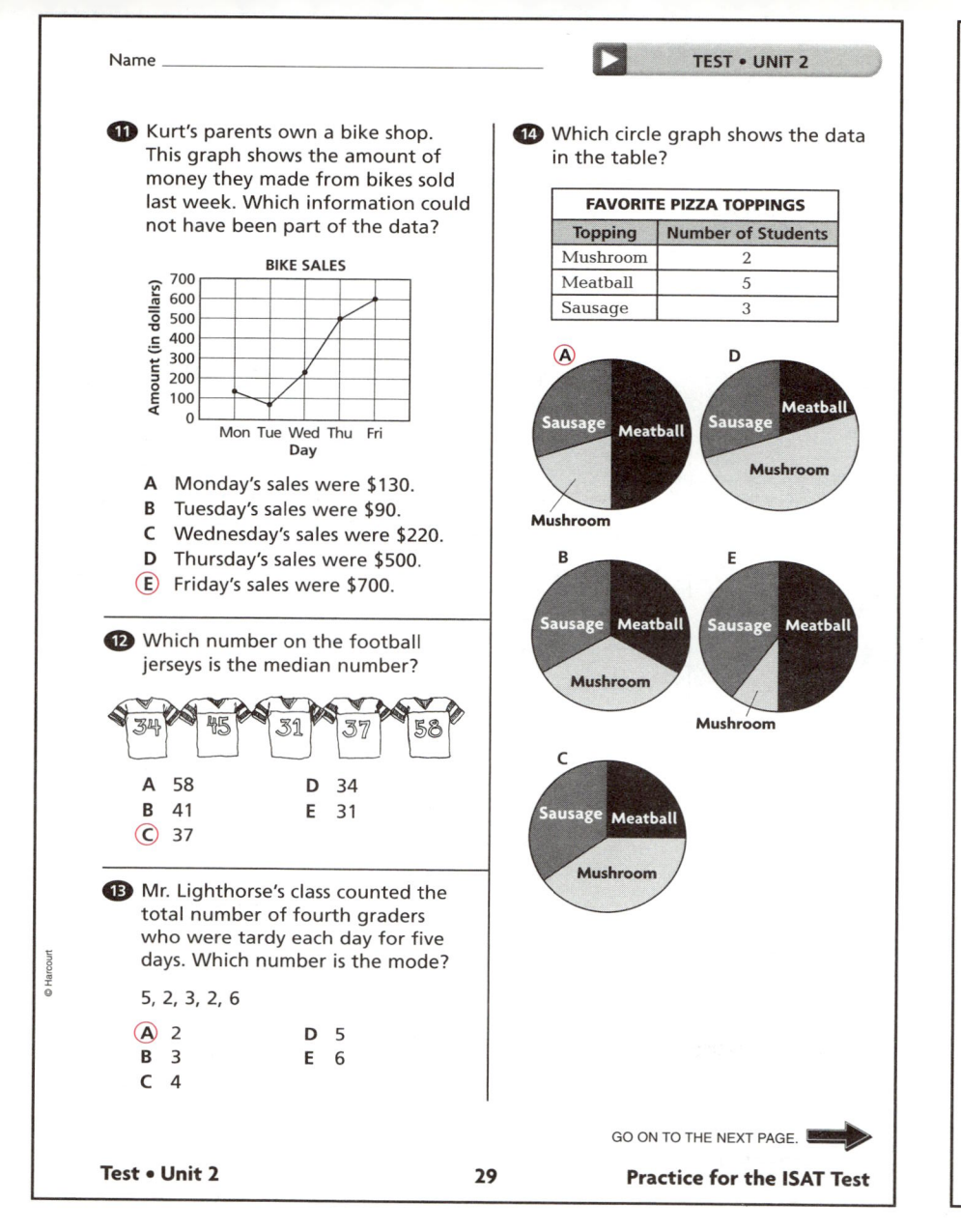

Name _____

11 Kurt's parents own a bike shop. This graph shows the amount of money they made from bikes sold last week. Which information could not have been part of the data?

BIKE SALES
(graph showing Mon–Fri)

A Monday's sales were $130.
B Tuesday's sales were $90.
C Wednesday's sales were $220.
D Thursday's sales were $500.
(E) Friday's sales were $700.

12 Which number on the football jerseys is the median number?

34 45 31 37 58

A 58   D 34
B 41   E 31
(C) 37

13 Mr. Lighthorse's class counted the total number of fourth graders who were tardy each day for five days. Which number is the mode?

5, 2, 3, 2, 6

(A) 2   D 5
B 3   E 6
C 4

14 Which circle graph shows the data in the table?

FAVORITE PIZZA TOPPINGS

Topping	Number of Students
Mushroom	2
Meatball	5
Sausage	3

(A), B, C, D, E (circle graphs)

Name _____

15 Winona wrote down how many minutes it took her to walk her dog each day for five days. Which is the median number of minutes she walked her dog?

Minutes Walked

Day	1	2	3	4	5
Minutes	22	15	18	25

A 15 minutes
B 18 minutes
C 21 minutes
(D) 22 minutes
E 25 minutes

16 Harlan took a survey of his class to determine the favorite swimming strokes of fourth graders. He made this double-bar graph to show the results. Which stroke did fewer boys than girls like?

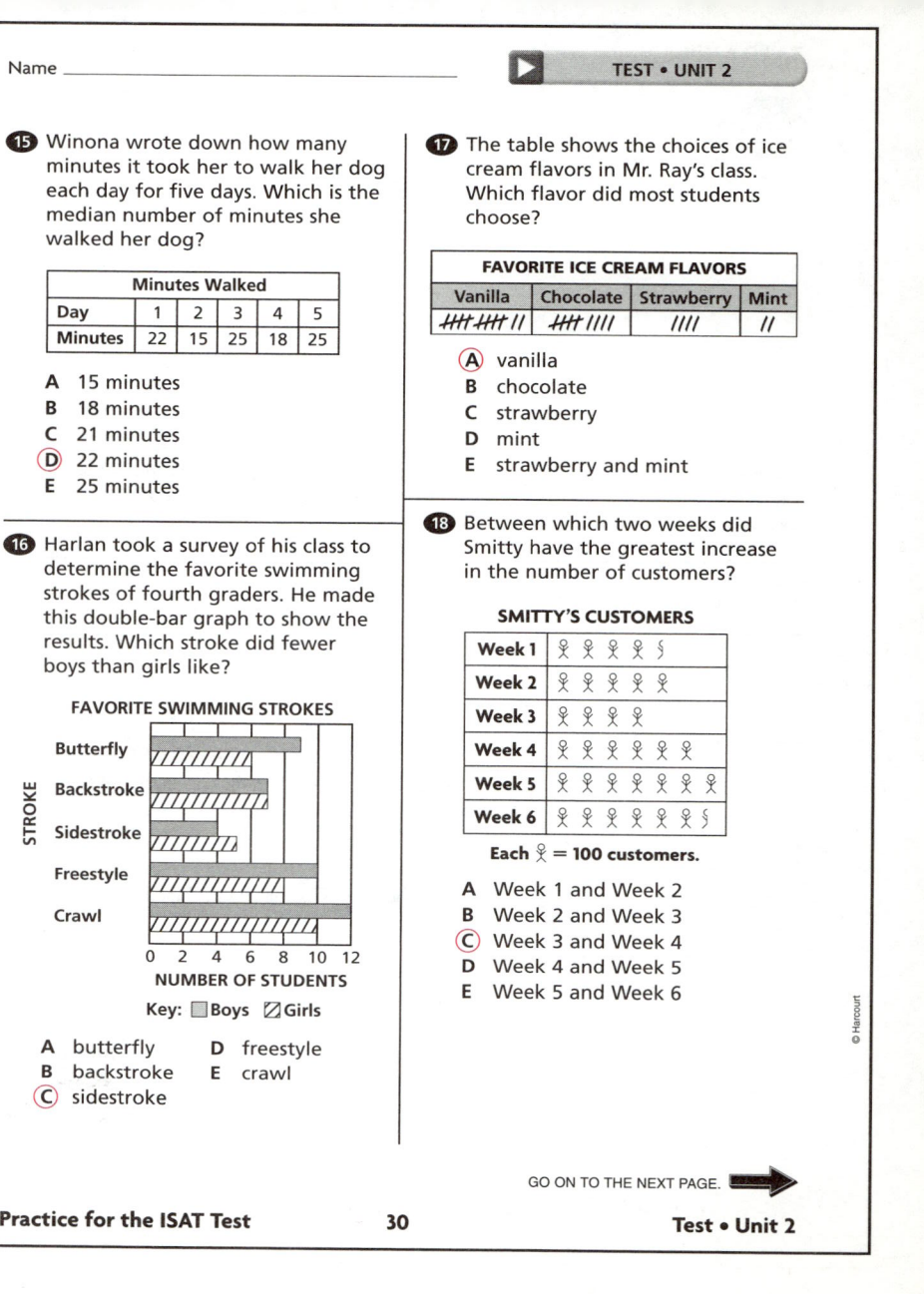

FAVORITE SWIMMING STROKES
(double-bar graph: Butterfly, Backstroke, Sidestroke, Freestyle, Crawl)
Key: Boys, Girls

A butterfly   D freestyle
B backstroke   E crawl
(C) sidestroke

17 The table shows the choices of ice cream flavors in Mr. Ray's class. Which flavor did most students choose?

FAVORITE ICE CREAM FLAVORS

Vanilla	Chocolate	Strawberry	Mint
### ### II	### IIII	IIII	II

(A) vanilla
B chocolate
C strawberry
D mint
E strawberry and mint

18 Between which two weeks did Smitty have the greatest increase in the number of customers?

SMITTY'S CUSTOMERS

Week 1	☻☻☻☻ 5
Week 2	☻☻☻
Week 3	☻☻☻
Week 4	☻☻☻☻☻
Week 5	☻☻☻☻☻☻
Week 6	☻☻☻☻☻

Each ☻ = 100 customers.

A Week 1 and Week 2
B Week 2 and Week 3
(C) Week 3 and Week 4
D Week 4 and Week 5
E Week 5 and Week 6

Name _____

19 An African elephant needs to drink up to 24 gallons of water each day. The table below shows how much water Daisy, an African elephant, drank over five days.

WATER DAISY DRANK

Day	Amount (in gallons)
Monday	24
Tuesday	19
Wednesday	13
Thursday	20
Friday	22

Use the information in the table to make a line graph. Label your graph.

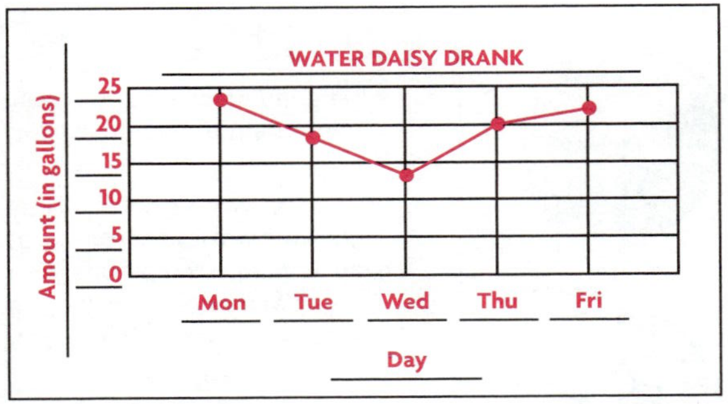

WATER DAISY DRANK

(line graph with points: Mon 24, Tue 19, Wed 13, Thu 20, Fri 22; y-axis Amount (in gallons) 0–25; x-axis Day)

Use your graph to answer the questions below.

Was the total amount that Daisy drank over five days greater than or less than 100 gallons?

Answer: _____less than_____

On which day did Daisy drink the least amount of water?

Answer: _____Wednesday_____

Test • Unit 2   31   Practice for the ISAT Test

GO ON TO THE NEXT PAGE.

Name _____

20 The pictograph below shows the average life spans of five animals.

AVERAGE LIFE SPANS OF SOME ANIMALS

Animal	Number of Years
Baboon	🟦🟦🟦🟦
Grizzly Bear	🟦🟦🟦
Cow	🟦🟦
Wolf	🟦
African Elephant	🟦🟦🟦🟦🟦🟦🟦🟦

Key: Each 🟦 = 5 years.

On the lines below, tell how you can find out how much longer a grizzly bear lives than a cow.

Possible answer: First, find out how long each animal lives. Each 🟦 equals 5 years, and a grizzly bear has 5 🟦, so it lives for 5 + 5 + 5 + 5 + 5 = 25 years. A cow has 3 🟦, so it lives for 5 + 5 + 5 = 15 years. To find out how much longer a grizzly bear lives than a cow, subtract: 25 − 15 = 10. A grizzly bear lives 10 years longer than a cow.

The average life span of an Asian elephant is about 25 years longer than the life span of a cow.

How many 🟦 would you need to show the life span of an Asian elephant on this graph?

Answer: ___8___

Practice for the ISAT Test   32   Test • Unit 2

STOP

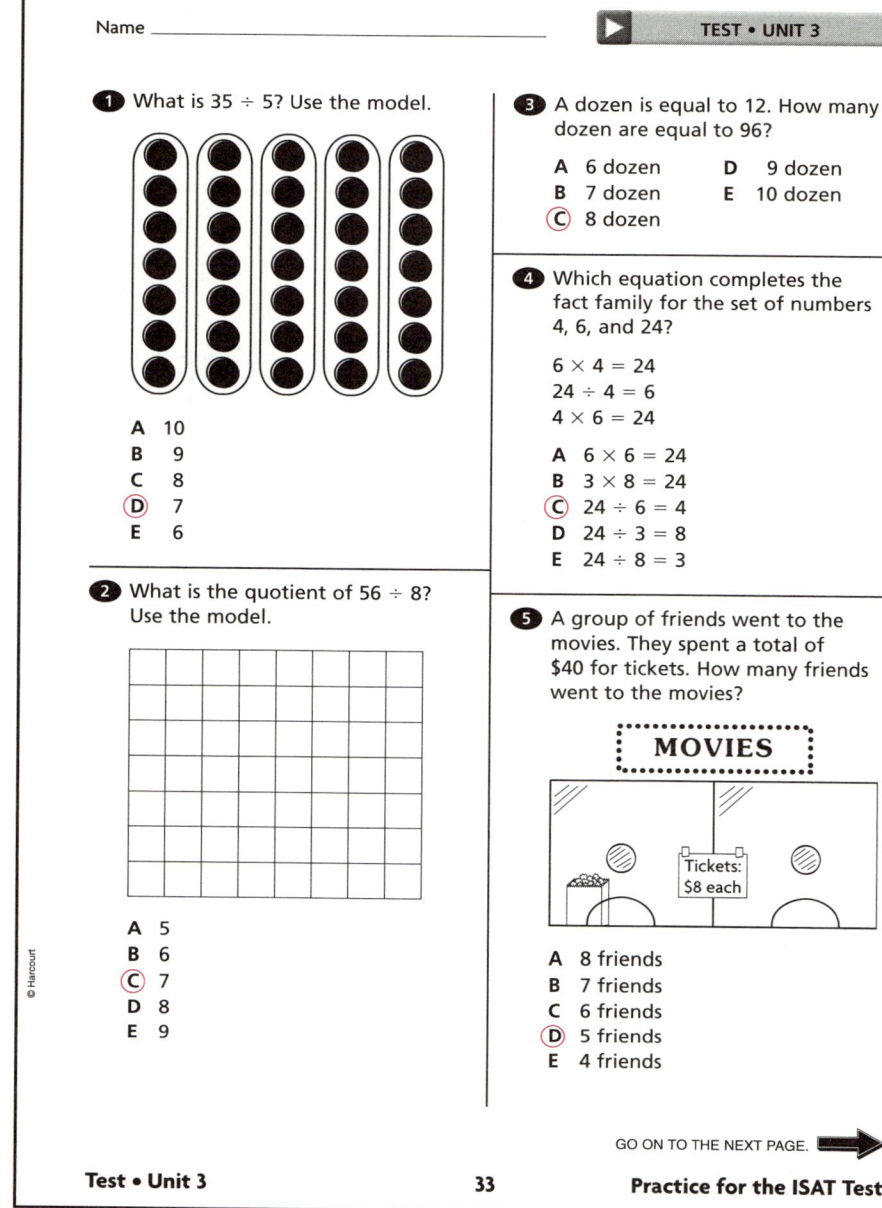

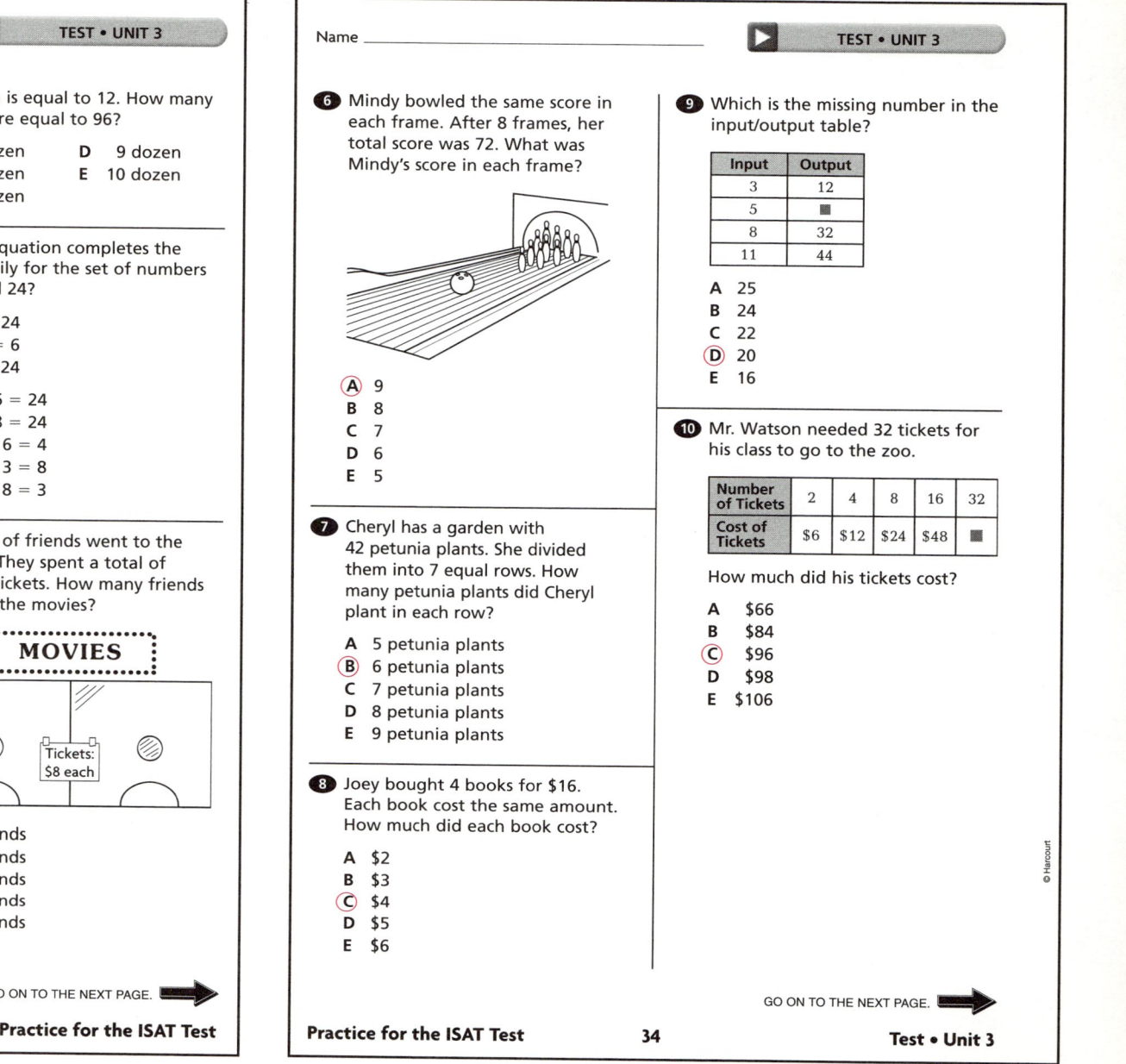

Name _____

TEST • UNIT 3

1 What is 35 ÷ 5? Use the model.

A 10
B 9
C 8
(D) 7
E 6

2 What is the quotient of 56 ÷ 8? Use the model.

A 5
B 6
(C) 7
D 8
E 9

3 A dozen is equal to 12. How many dozen are equal to 96?

A 6 dozen   D 9 dozen
B 7 dozen   E 10 dozen
(C) 8 dozen

4 Which equation completes the fact family for the set of numbers 4, 6, and 24?

6 × 4 = 24
24 ÷ 4 = 6
4 × 6 = 24

A 6 × 6 = 24
B 3 × 8 = 24
(C) 24 ÷ 6 = 4
D 24 ÷ 3 = 8
E 24 ÷ 8 = 3

5 A group of friends went to the movies. They spent a total of $40 for tickets. How many friends went to the movies?

Tickets: $8 each

A 8 friends
B 7 friends
C 6 friends
(D) 5 friends
E 4 friends

Test • Unit 3   33   Practice for the ISAT Test

GO ON TO THE NEXT PAGE.

TEST • UNIT 3

6 Mindy bowled the same score in each frame. After 8 frames, her total score was 72. What was Mindy's score in each frame?

(A) 9
B 8
C 7
D 6
E 5

7 Cheryl has a garden with 42 petunia plants. She divided them into 7 equal rows. How many petunia plants did Cheryl plant in each row?

A 5 petunia plants
(B) 6 petunia plants
C 7 petunia plants
D 8 petunia plants
E 9 petunia plants

8 Joey bought 4 books for $16. Each book cost the same amount. How much did each book cost?

A $2
B $3
(C) $4
D $5
E $6

9 Which is the missing number in the input/output table?

Input	Output
3	12
5	■
8	32
11	44

A 25
B 24
C 22
(D) 20
E 16

10 Mr. Watson needed 32 tickets for his class to go to the zoo.

Number of Tickets	2	4	8	16	32
Cost of Tickets	$6	$12	$24	$48	■

How much did his tickets cost?

A $66
B $84
(C) $96
D $98
E $106

Practice for the ISAT Test   34   Test • Unit 3

GO ON TO THE NEXT PAGE.

Name _____

11 Which value for the variable makes the equation true?

$7 \times n = 77$

A $n = 539$
B $n = 84$
C $n = 70$
D $n = 11$
E $n = 10$

12 Which value for the variable makes the equation true?

$n \times 8 = 72$

A $n = 11$
B $n = 9$
C $n = 8$
D $n = 7$
E $n = 5$

13 Which operation should you do first?

$5 + (8 \times 4) + 16$

A $5 + 8$
B $5 + 4$
C $5 + 16$
D $4 + 16$
E 8×4

14 Luci wrote an expression on the board.

$16 \times 4 \times 5$

Then Luci explained how she got her answer. "First I multiplied to get 20, and then I multiplied that by 16 to get 320."

Which expression shows how Luci got her answer?

A $16 + (4 \times 5)$
B $16 \times (4 \times 5)$
C $(5 \times 16) \times 4$
D $5 \times (4 \times 16)$
E $(16 \times 4) \times 5$

15 Which is the value of the expression?

$(37 \times 5) \times 2$

A 372
B 370
C 360
D 335
E 307

Name _____

16 Which is the value of the expression?

$2 \times (7 + 8)$

A 112
B 60
C 30
D 22
E 17

17 Sammie described a number pattern that she created. The inputs (x) are 54, 48, 42, and 36. The related outputs (y) are 9, 8, 7, and 6.

Which equation shows the rule for Sammie's pattern?

A $x - 6 = y$
B $x + y = 6$
C $x \times 6 = y$
D $x \div 6 = y$
E $6 \div x = y$

18 Which pair of related equations is represented by the model below?

A $3 + 4 = 7$ and $7 - 4 = 3$
B $4 \times 4 = 16$ and $16 \div 4 = 4$
C $3 \times 4 = 12$ and $16 \div 4 = 4$
D $3 \times 4 = 12$ and $12 \div 3 = 4$
E $4 \times 4 = 16$ and $12 \div 3 = 4$

19 What is $48 \div 8$?

A 9
B 8
C 7
D 6
E 5

20 What is $36 \div 9$?

A 16
B 15
C 4
D 3
E 2

Name _____

21 Lori is buying juice boxes for her party. She is having 28 guests.

JUICE BOXES		
Package Size	Boxes per Package	Cost per Package
Medium	6	$2

If Lori buys 4 medium packages of juice boxes, will she have enough so that each guest will have a box of juice?

On the lines below, tell which operation is needed to solve the problem. Explain why. Then solve.

Possible explanation: Multiplication, because I need to find out how many boxes of juice there are in 4 medium packages. Then I can compare that amount to see whether it is greater than or less than 28.

Answer:
4 × 6 = 24; 24 < 28, so there are not enough juice boxes for all the guests.

Lori decides to spend $10 on juice boxes. How many medium packages does she buy? How many juice boxes does she buy?

Show your work.

$10 ÷ $2 = 5 packages
5 × 6 = 30 juice boxes

Answer: __5__ packages
__30__ juice boxes

Name _____

22 The map shows the distances in miles between four places.

[Map: School — 5 miles — Park; School — 3 miles — Marvin's House; Marvin's House — 1 mile — Library; Library — 2 miles — Park]

Marvin rode his bike from his house to school and then back home each of the 5 days of the school week. On Friday, he also rode from his house to the library and back home. How many miles did he ride in all?

Write an expression to match the words. Then find the value of the expression.

Answer:
Possible answer: (6 × 5) + 2; 32 miles

Sophie rode from school to the park and then back to school. Then she rode to Marvin's house. How many miles did she ride in all?

Write an expression to match the words. Find the value of the expression.

Answer:
Possible answer: (5 × 2) + 3; 13 miles

TEST • UNIT 4

Name _____

1 Tim and his dad have been to 7 Chicago Bulls basketball games during the past season. They live 32 miles from the arena. Tim figured out that they traveled 448 miles in all. Is that reasonable?

- A No. 7 × 32 is about 7 × 30, or 210 miles.
- B No. Tim is just guessing. There is no way to know how far they traveled.
- C No. The answer should be about 320 miles.
- (D) Yes. A round trip is about 60 miles and 7 × 60 = 420 miles.
- E No. 448 miles seems too many because the arena is only 32 miles away.

2 Last year, each student and teacher at Lakeview School used an average of 18 pounds of paper. There are a total of 314 students and teachers in the school. Estimate how much paper the school used last year.

- A about 300 pounds
- B about 600 pounds
- C about 3,000 pounds
- (D) about 6,000 pounds
- E about 8,000 pounds

3 Amy counted 20 flowers in one section of her flower garden. About how many flowers are in the garden in all?

- A 20 flowers
- B 40 flowers
- (C) 100 flowers
- D 200 flowers
- E 300 flowers

4 At the Carr Memorial Theatre, 995 seats are arranged in 48 rows. Using a basic fact and a pattern, which is the most reasonable estimate for the number of seats in each row?

- A 2 seats
- B 3 seats
- C 10 seats
- (D) 20 seats
- E 30 seats

5 Which estimate would you use to check this expression?

324 × 47

- A 320 × 40 = 12,800
- B 330 × 40 = 13,200
- (C) 320 × 50 = 16,000
- D 330 × 50 = 16,500
- E 330 × 40 = 13,200

GO ON TO THE NEXT PAGE.

Test • Unit 4   39   Practice for the ISAT Test

TEST • UNIT 4

Name _____

6 Grills Unlimited sold 9 Barbeque King grills last week. How much money did Grills Unlimited earn from selling these grills?

Barbeque King $527

- A $4,553
- B $4,683
- (C) $4,743
- D $5,279
- E $5,323

7 Look at the pattern below. What is the value of n?

$5 \times 8 = 40$
$5 \times 80 = 400$
$5 \times n = 4,000$

- A $n = 8,000$
- (B) $n = 800$
- C $n = 88$
- D $n = 80$
- E $n = 8$

8 Tom is building 4 new bookcases for the library. He needs 48 screws for each bookcase. How many screws does he need for all 4 bookcases?

- A 162 screws
- B 168 screws
- (C) 192 screws
- D 208 screws
- E 218 screws

9 What is the value of n in this equation?

$9 \times n = 4,500$

- A $n = 5$
- B $n = 50$
- (C) $n = 500$
- D $n = 4,000$
- E $n = 5,000$

10 Jess prints and sells T-shirts for $8 each. How much would Jess make if she sold 281 T-shirts?

- A $1,648
- B $1,668
- C $2,068
- (D) $2,248
- E $2,648

GO ON TO THE NEXT PAGE.

Practice for the ISAT Test   40   Test • Unit 4

Name _____

11 12 × 11 is more than 11 × 11. How much more?
- (A) 11
- B 12
- C 111
- D 121
- E 144

12 Rachel is going on a walk to raise money for a local charity. Her parents will donate $2 for the first mile she walks, $4 for the second mile, and an additional $2 for each mile after that. How much money will Rachel raise if she walks 7 miles?
- A $11
- B $14
- C $42
- (D) $56
- E $72

13 Twelve pine trees are growing in 1 acre of a forest. How many pine trees are in 9 acres?
- A 21 pine trees
- B 84 pine trees
- C 96 pine trees
- D 98 pine trees
- (E) 108 pine trees

14 Angela is planning a garden. She will plant 9 rows of flowers. Each row will have 10 tulips and 5 daisies. How many flowers will there be in all?
- A 45 flowers
- B 90 flowers
- (C) 135 flowers
- D 180 flowers
- E 450 flowers

15 An outdoor park had 12 visitors one day during the summer. If the ticket price was $8, how much money in ticket sales was made that day?
- A $20
- B $72
- C $84
- D $86
- (E) $96

Test • Unit 4   41   Practice for the ISAT Test

Name _____

16 The table shows the amount of money Chef's Pantry pays Clay for every 2 potholders he makes. If he makes 18 potholders, how much will he earn?

POTHOLDERS CLAY MAKES

Number of Potholders	Money Earned
2	$8
4	$16
6	$24
10	$40
12	$48

- A $56
- B $64
- (C) $72
- D $80
- E $88

17 A factory makes 70 rubber balls each hour. How many balls can the factory make in one 40-hour work week?
- A 24 balls
- B 28 balls
- C 280 balls
- (D) 2,800 balls
- E 28,000 balls

18 Mr. and Mrs. Anderson and their 3 children are taking a train trip to visit their relatives. Each ticket costs $80. What is the total cost of the tickets?
- (A) $400
- B $320
- C $240
- D $160
- E $110

19 An 8-ounce container of strawberry yogurt has 230 calories. How many calories are there in 3 containers?

Write a number sentence, and then solve the problem.

- A 8 × 230 = n; 1,840 calories
- B 8 × 230 = n; 1,640 calories
- (C) 3 × 230 = n; 690 calories
- D 3 × 230 = n; 590 calories
- E 8 × 230 × 3 = n; 5,520 calories

20 Marissa keeps her CDs in a box. Each box holds 30 CDs.

Number of Boxes	1	2	3	4	5	6
CDs in Each Box	30	60	90			

Which set of numbers completes the table?
- A 91, 92, 93
- B 30, 60, 90
- (C) 120, 150, 180
- D 160, 200, 240
- E 400, 450, 500

Practice for the ISAT Test   42   Test • Unit 4

Name _____

21 Use place value and expanded notation to find the product 4 × 276.

276 in expanded form is 200 + 70 + 6.

Now, multiply. Then add.

4 × 276

(4 × 200) + (4 × 70) + (4 × 6)

   800   +   280   +   24   =   1,104

Use the method above and find the product 5 × 839.
Show your work.

(5 × 800) + (5 × 30) + (5 × 9)

  4,000 +   150   +   45   =   4,195

GO ON TO THE NEXT PAGE.

Name _____

22 The city of Maplewood Park is planning new gardens for its city parks. The city's landscape designer drew this garden design.

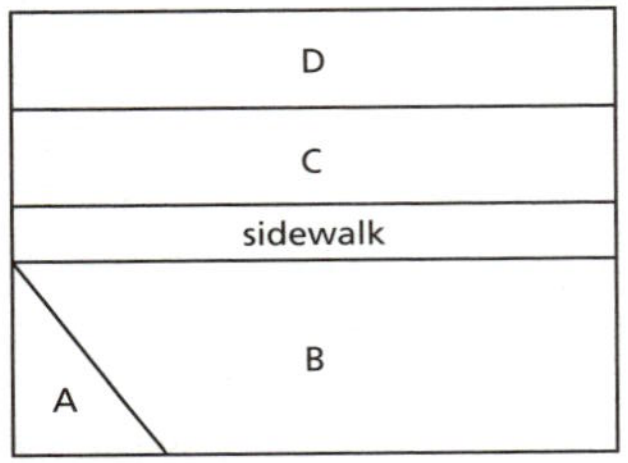

A	D
	C
	sidewalk
	B

In section A, the designer plans to plant 77 coneflowers. What would be a good estimate for the number of coneflowers that could be planted in section B? Explain.
Possible answer: 560 coneflowers; Section B is about 7 times as large as section A. 7 × 80 = 560, so about 560 coneflowers.

In section D, the designer plans to plant 8 rows of 34 dahlias each. How many dahlias does the designer plan to plant?
272 dahlias

In section C, the designer plans to plant 9 rows of 25 roses each. How many roses does the designer plan to plant?
225 roses

STOP

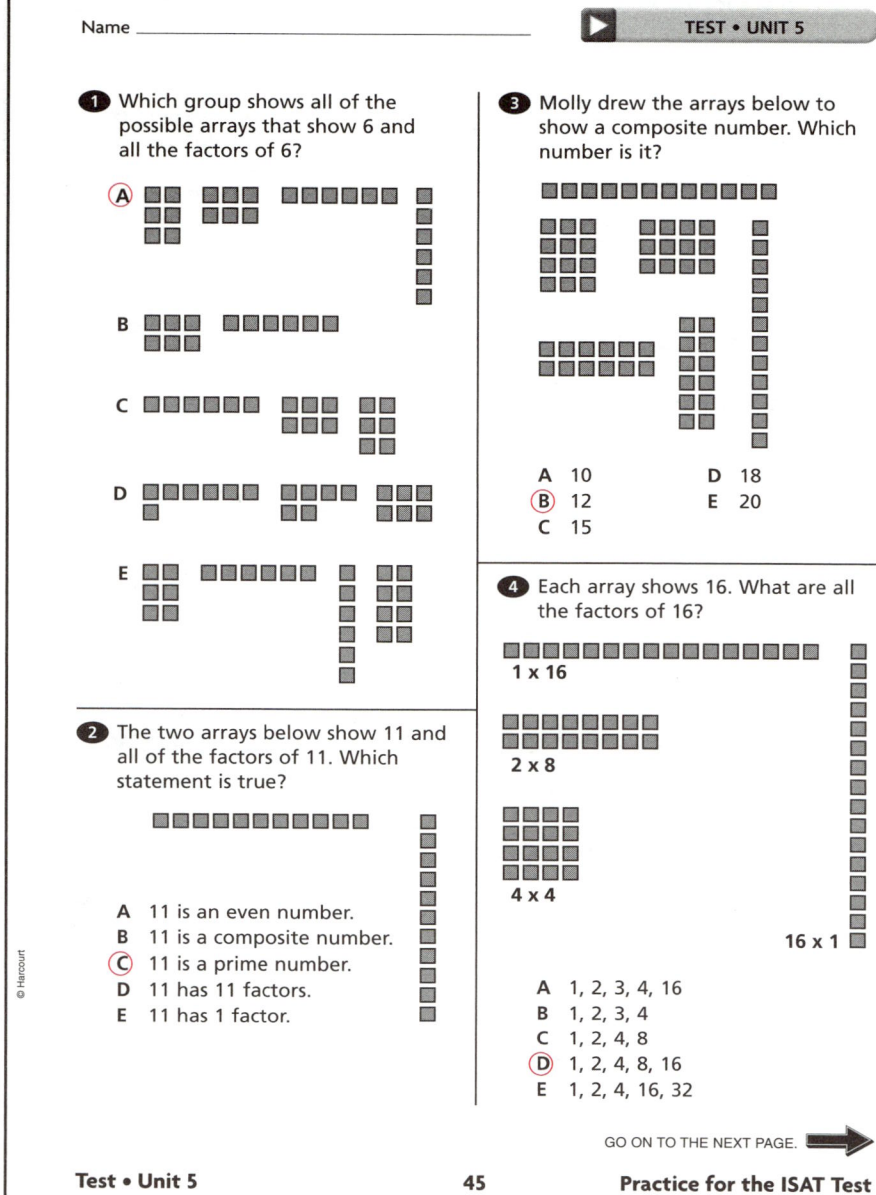

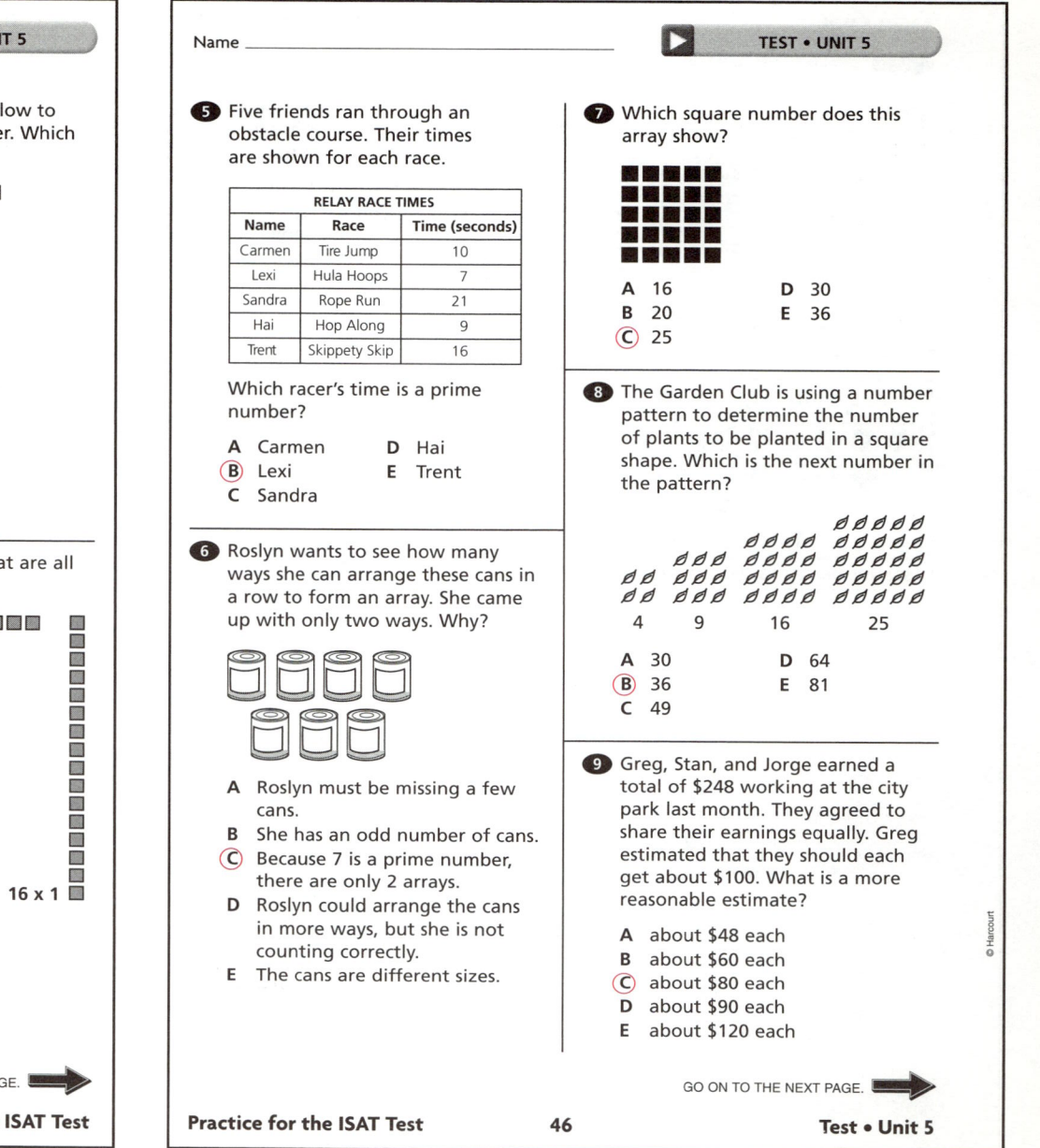

Name _____

TEST • UNIT 5

1 Which group shows all of the possible arrays that show 6 and all the factors of 6?

(A)
B
C
D
E

2 The two arrays below show 11 and all of the factors of 11. Which statement is true?

A 11 is an even number.
B 11 is a composite number.
(C) 11 is a prime number.
D 11 has 11 factors.
E 11 has 1 factor.

3 Molly drew the arrays below to show a composite number. Which number is it?

A 10
(B) 12
C 15
D 18
E 20

4 Each array shows 16. What are all the factors of 16?

1 × 16
2 × 8
4 × 4
16 × 1

A 1, 2, 3, 4, 16
B 1, 2, 3, 4
C 1, 2, 4, 8
(D) 1, 2, 4, 8, 16
E 1, 2, 4, 16, 32

Test • Unit 5   45   Practice for the ISAT Test

GO ON TO THE NEXT PAGE.

Name _____

TEST • UNIT 5

5 Five friends ran through an obstacle course. Their times are shown for each race.

RELAY RACE TIMES		
Name	Race	Time (seconds)
Carmen	Tire Jump	10
Lexi	Hula Hoops	7
Sandra	Rope Run	21
Hai	Hop Along	9
Trent	Skippety Skip	16

Which racer's time is a prime number?

A Carmen   D Hai
(B) Lexi   E Trent
C Sandra

6 Roslyn wants to see how many ways she can arrange these cans in a row to form an array. She came up with only two ways. Why?

A Roslyn must be missing a few cans.
B She has an odd number of cans.
(C) Because 7 is a prime number, there are only 2 arrays.
D Roslyn could arrange the cans in more ways, but she is not counting correctly.
E The cans are different sizes.

7 Which square number does this array show?

A 16
B 20
(C) 25
D 30
E 36

8 The Garden Club is using a number pattern to determine the number of plants to be planted in a square shape. Which is the next number in the pattern?

4   9   16   25

A 30
(B) 36
C 49
D 64
E 81

9 Greg, Stan, and Jorge earned a total of $248 working at the city park last month. They agreed to share their earnings equally. Greg estimated that they should each get about $100. What is a more reasonable estimate?

A about $48 each
B about $60 each
(C) about $80 each
D about $90 each
E about $120 each

Practice for the ISAT Test   46   Test • Unit 5

GO ON TO THE NEXT PAGE.

TEST • UNIT 5

10 Steve does not want his uniform number to be a square number. Which number can he wear?

A 4   D 25
B 9   E 49
(C) 12

11 Carver Elementary School has 524 students. All of them are going on a trip to Washington, D.C. The school has rented 11 buses for the trip. About how many students will travel on each bus?

A about 40 students
(B) about 50 students
C about 60 students
D about 70 students
E about 80 students

12 Monica's grandmother lives 504 miles away. If the average speed of travel is 55 miles per hour, about how many hours will it take Monica's family to drive to her grandmother's house?

(A) 9 hours   D 60 hours
B 12 hours   E 65 hours
C 30 hours

13 King Elementary School has 7 fourth-grade classes. Each class has 27 students. How many fourth grade students go to King Elementary?

A 234 students
B 209 students
(C) 189 students
D 169 students
E 149 students

14 It took Greta a total of 109 minutes to do her homework for 5 different subjects. She worked on each subject for about the same amount of time. About how long did she work on each subject?

A between 5 and 10 minutes each
B between 10 and 20 minutes each
(C) between 20 and 30 minutes each
D between 30 and 40 minutes each
E between 40 and 45 minutes each

15 Which division equation is represented by the model?

A $128 \div 42 = 3\ r2$
(B) $129 \div 42 = 3\ r3$
C $129 \div 40 = 3\ r3$
D $126 \div 42 = 3\ r1$
E $126 \div 40 = 3$

16 Which is a related division sentence for $7 \times 8 = 56$?

(A) $56 \div 7 = 8$
B $56 \div 6 = 8$
C $49 \div 7 = 7$
D $48 \div 8 = 6$
E $42 \div 7 = 6$

TEST • UNIT 5

17 Ellen drove 325 miles to work and back last week. If she drove the same number of miles on each of 5 days, how many miles did she drive per day? Use the model.

A 61 miles
(B) 65 miles
C 71 miles
D 75 miles
E 81 miles

18 A beluga whale in captivity eats about 511 pounds of fish per week. How many pounds per day is that?

(A) 73 pounds
B 102 pounds
C 504 pounds
D 518 pounds
E 703 pounds

19 The number of seats in the school auditorium can be described by the equation $35 \times n = 315$. If n represents the number of rows, how many rows are there?

A 7 rows   D 10 rows
B 8 rows   E 11 rows
(C) 9 rows

20 Tad can write his full name in 5 seconds. He wrote the following equation to find out how many times he can write his name in 60 seconds.

$60 \div 5 = n$

Solve the equation for n.

A $n = 6$
(B) $n = 12$
C $n = 15$
D $n = 55$
E $n = 65$

21 Jenna bought flowers to create 4 table arrangements for a party. The pictograph shows how many of each type of flower she bought.

FLOWERS FOR TABLE ARRANGEMENTS

Tulips	❁❁
Daisies	❁
Tiger Lilies	❁❁❁
Daffodils	❁❁❁

Key: Each ❁ = 5 flowers.

Each arrangement has the same number of flowers. How many flowers are in each arrangement?

A 10 flowers
B 15 flowers
(C) 20 flowers
D 25 flowers
E 30 flowers

Name _____

22 The Glendale Marching Band has 20 members. The leader wants the band to form rows with 3 members in each row. How many rows will there be? How many members will be left over?

Draw a model showing counters to find out how many equal groups of 3 are in 20 and how many are left over.

(model showing 6 circles with 3 counters each, plus 2 counters left over)

How many rows of 3 can the band members form?
How many band members are left over?

Answer: __6__ rows with __2__ band members left over

Name _____

23 All 78 fourth-grade students from Winslow School went on a trip to the zoo. The zookeeper put students in groups of 3 to 5 students. All of the groups were the same size, and no students were left over. How many students were in each group? How many groups were there?

On the lines below, explain how you can use the strategy predict and test to solve this problem.

Possible explanation: Each group can have 3, 4, or 5 students. So, I predicted that either 3, 4, or 5 would be correct. To test, I can divide 78 by each number and see which division has no remainder.

Use the strategy predict and test to solve the problem.
Show all work.

**Possible answer: Use 3, 4, and 5 as divisors.
78 ÷ 3 = 26; 78 ÷ 4 = 19 r2; 78 ÷ 5 = 15 r3.
If 3 students are in each group, no students will be left over.**

Answer: __3__ students in each group and __26__ groups

Name _____

TEST • UNIT 6

1. The record high temperature in the desert is 118°F. Which thermometer shows that temperature?

   A, B, C (marked), D, E

2. Marti's refrigerator is set at 40°F. Which thermometer shows that temperature?

   A, B, C (marked), D, E

3. Find the change in temperature from 22°C to 35°C.
   A (marked) 13°C
   B 15°C
   C 17°C
   D 22°C
   E 57°C

4. Which is not a regular polygon?

   A, B (marked), C, D, E

5. A honeycomb is made of cells in the shape of hexagons. Which statement is not true?

   A A hexagon has 6 sides.
   B A hexagon has 6 lines of symmetry.
   C The sides of a regular hexagon have the same length.
   D (marked) A regular hexagon has the same shape as an octagon.
   E The diagram shows a regular hexagon.

6. Ms. Lynch wants to design her kitchen floor with polygons. Which shape could not be used?

   A (marked) circle
   B triangle
   C square
   D pentagon
   E heptagon

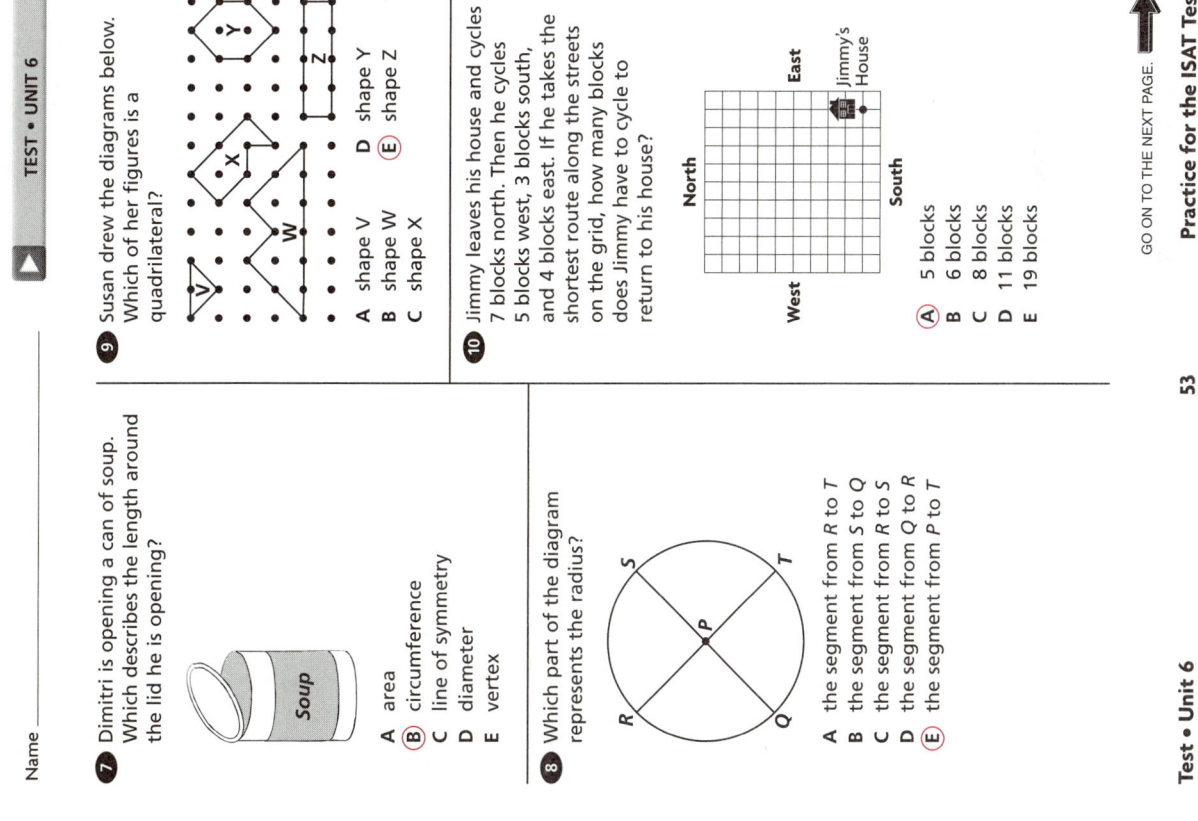

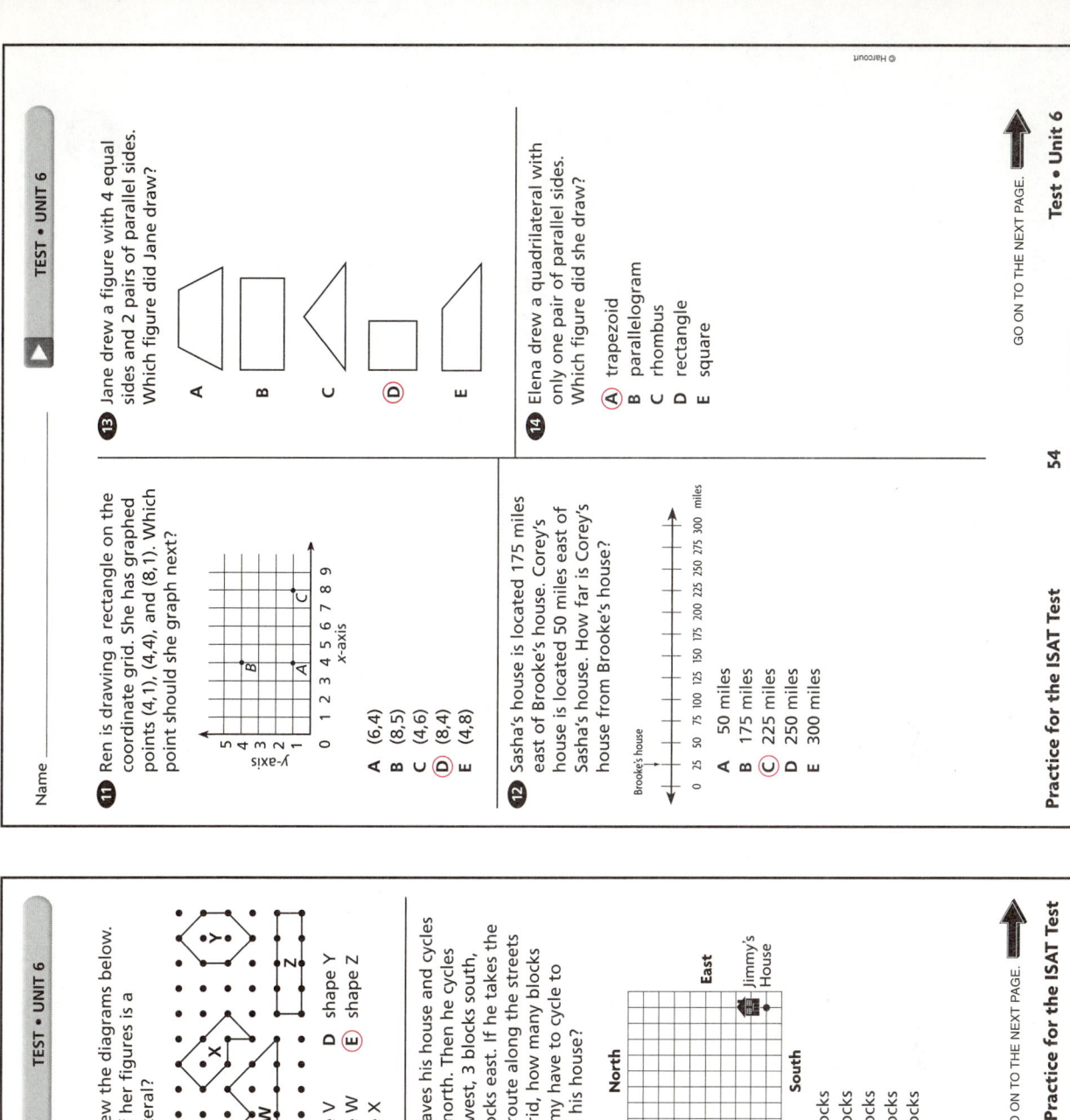

Name _____

7 Dimitri is opening a can of soup. Which describes the length around the lid he is opening?

- A area
- (B) circumference
- C line of symmetry
- D diameter
- E vertex

8 Which part of the diagram represents the radius?

- A the segment from R to T
- B the segment from S to Q
- C the segment from R to S
- D the segment from Q to R
- (E) the segment from P to T

9 Susan drew the diagrams below. Which of her figures is a quadrilateral?

- A shape V
- B shape W
- C shape X
- D shape Y
- (E) shape Z

10 Jimmy leaves his house and cycles 7 blocks north. Then he cycles 5 blocks west, 3 blocks south, and 4 blocks east. If he takes the shortest route along the streets on the grid, how many blocks does Jimmy have to cycle to return to his house?

- (A) 5 blocks
- B 6 blocks
- C 8 blocks
- D 11 blocks
- E 19 blocks

Test • Unit 6   53   Practice for the ISAT Test

Name _____

11 Ren is drawing a rectangle on the coordinate grid. She has graphed points (4,1), (4,4), and (8,1). Which point should she graph next?

- A (6,4)
- B (8,5)
- C (4,6)
- (D) (8,4)
- E (4,8)

12 Sasha's house is located 175 miles east of Brooke's house. Corey's house is located 50 miles east of Sasha's house. How far is Corey's house from Brooke's house?

- A 50 miles
- B 175 miles
- (C) 225 miles
- D 250 miles
- E 300 miles

13 Jane drew a figure with 4 equal sides and 2 pairs of parallel sides. Which figure did Jane draw?

- A
- B
- C
- (D)
- E

14 Elena drew a quadrilateral with only one pair of parallel sides. Which figure did she draw?

- (A) trapezoid
- B parallelogram
- C rhombus
- D rectangle
- E square

Practice for the ISAT Test   54   Test • Unit 6

Name _____

TEST • UNIT 6

15 These figures are all quadrilaterals.

Sort the figures into the diagram below.

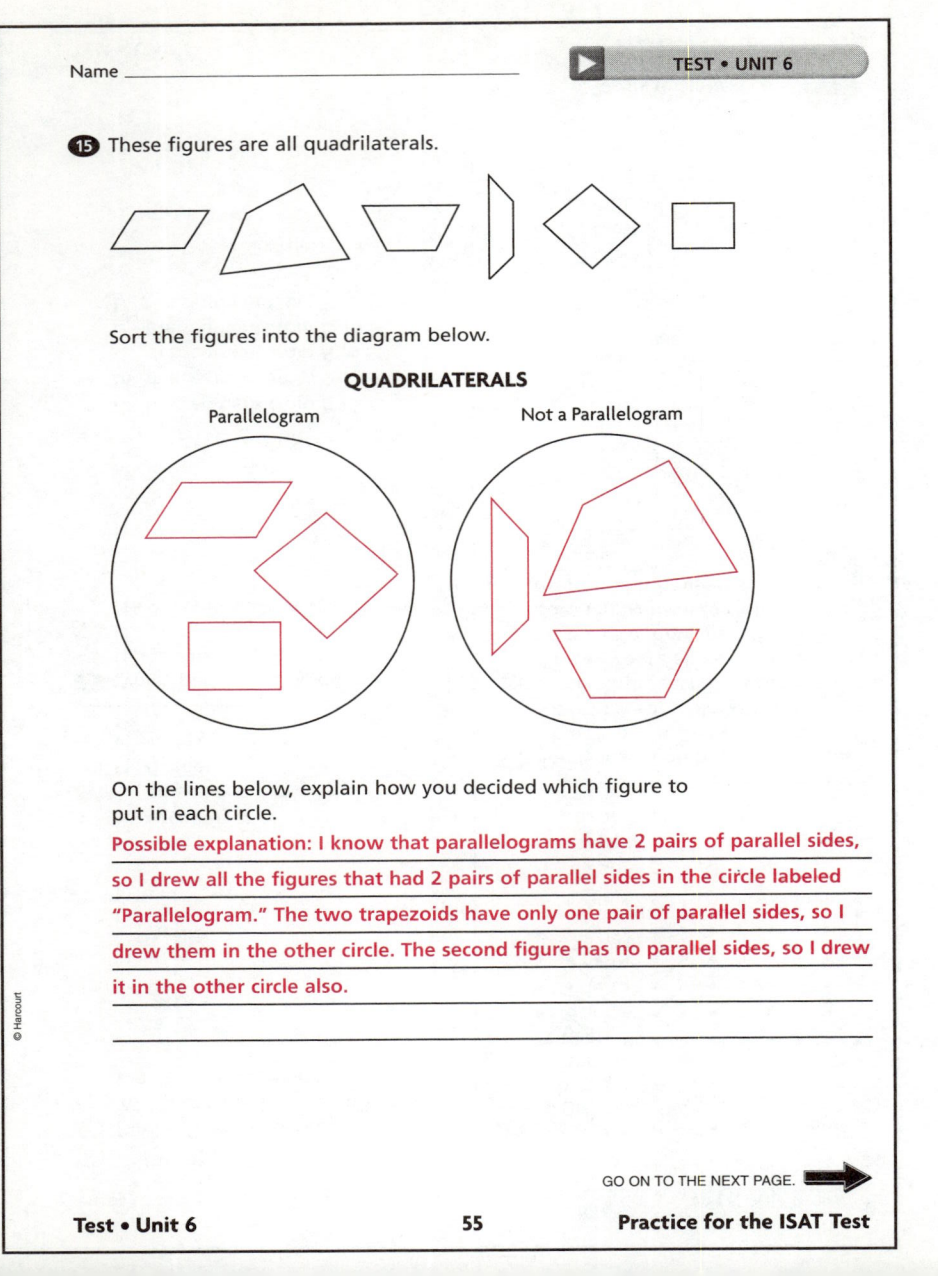

QUADRILATERALS

Parallelogram | Not a Parallelogram

On the lines below, explain how you decided which figure to put in each circle.

Possible explanation: I know that parallelograms have 2 pairs of parallel sides, so I drew all the figures that had 2 pairs of parallel sides in the circle labeled "Parallelogram." The two trapezoids have only one pair of parallel sides, so I drew them in the other circle. The second figure has no parallel sides, so I drew it in the other circle also.

Test • Unit 6   55   Practice for the ISAT Test

GO ON TO THE NEXT PAGE.

Name _____

TEST • UNIT 6

16 Leisha is designing a quilted seat pad for her kitchen chairs. After she sews the pieces together, she will place yarn knots along the outline of her design. Graph and label each ordered pair below to see where she plans to place the yarn knots. Possible answers are given.

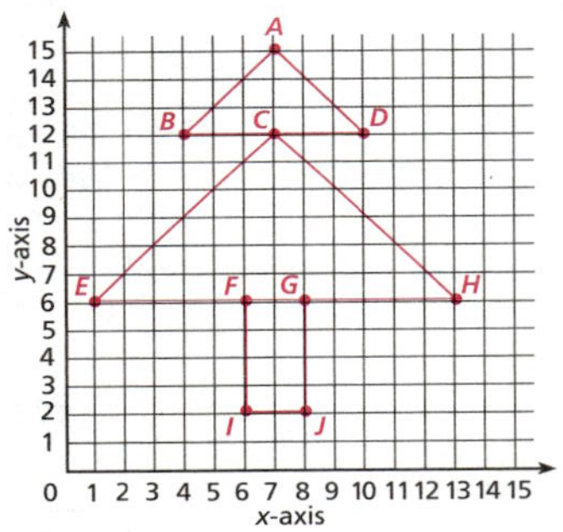

Point	Ordered Pair
A	(7,15)
B	(4,12)
C	(7,12)
D	(10,12)
E	(1,6)
F	(6,6)
G	(8,6)
H	(13,6)
I	(6,2)
J	(8,2)

Connect points A, B, and D. Name the polygon. triangle or isosceles triangle
Connect points C, E, and H. Name the polygon. triangle or isosceles triangle
Connect points F, G, J, and I. Name the polygon. rectangle or parallelogram
What design did Leisha quilt? a tree

Use the coordinate grid below. Make a quilt design by connecting ordered pairs. Graph and label the ordered pairs. Then identify the polygons in your design. Check students' work.

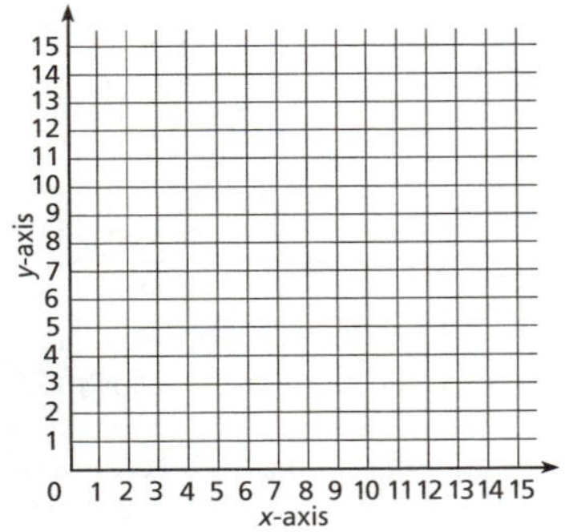

Practice for the ISAT Test   56   Test • Unit 6

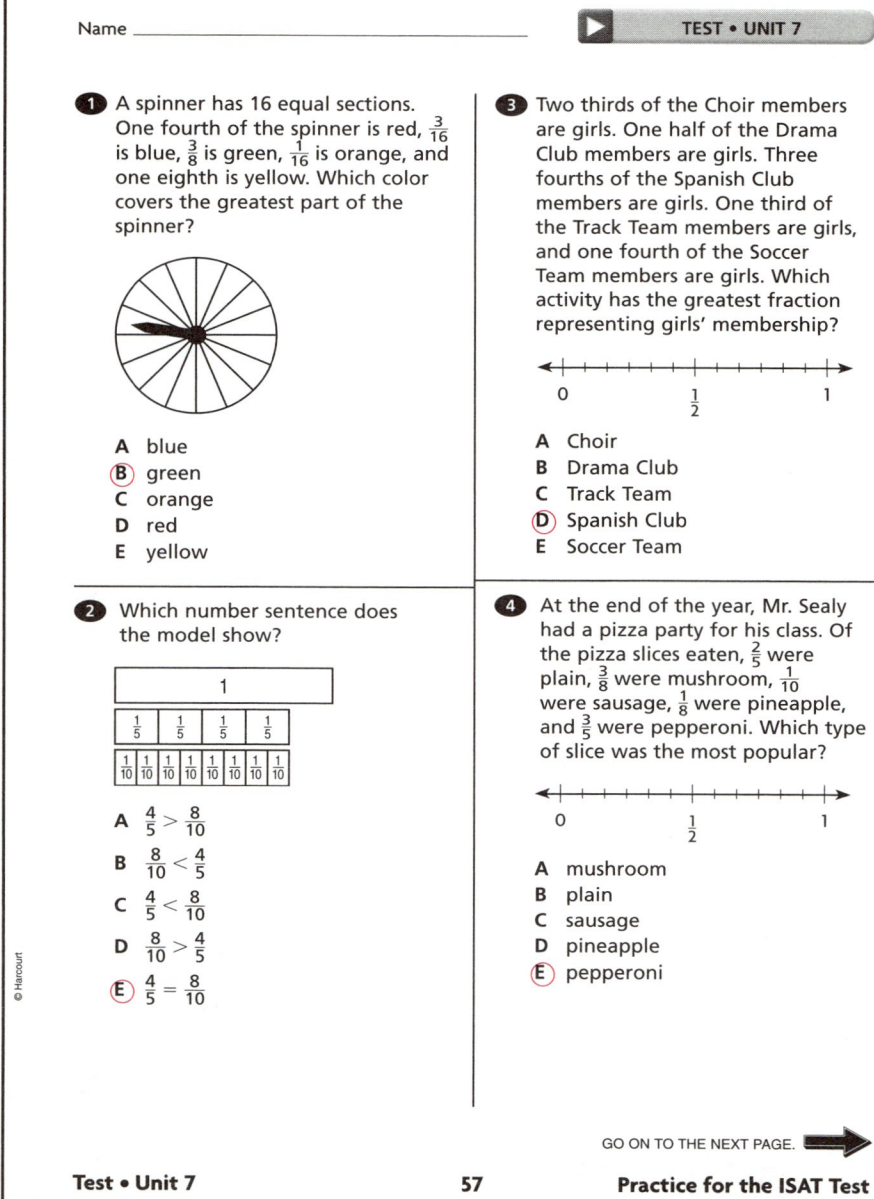

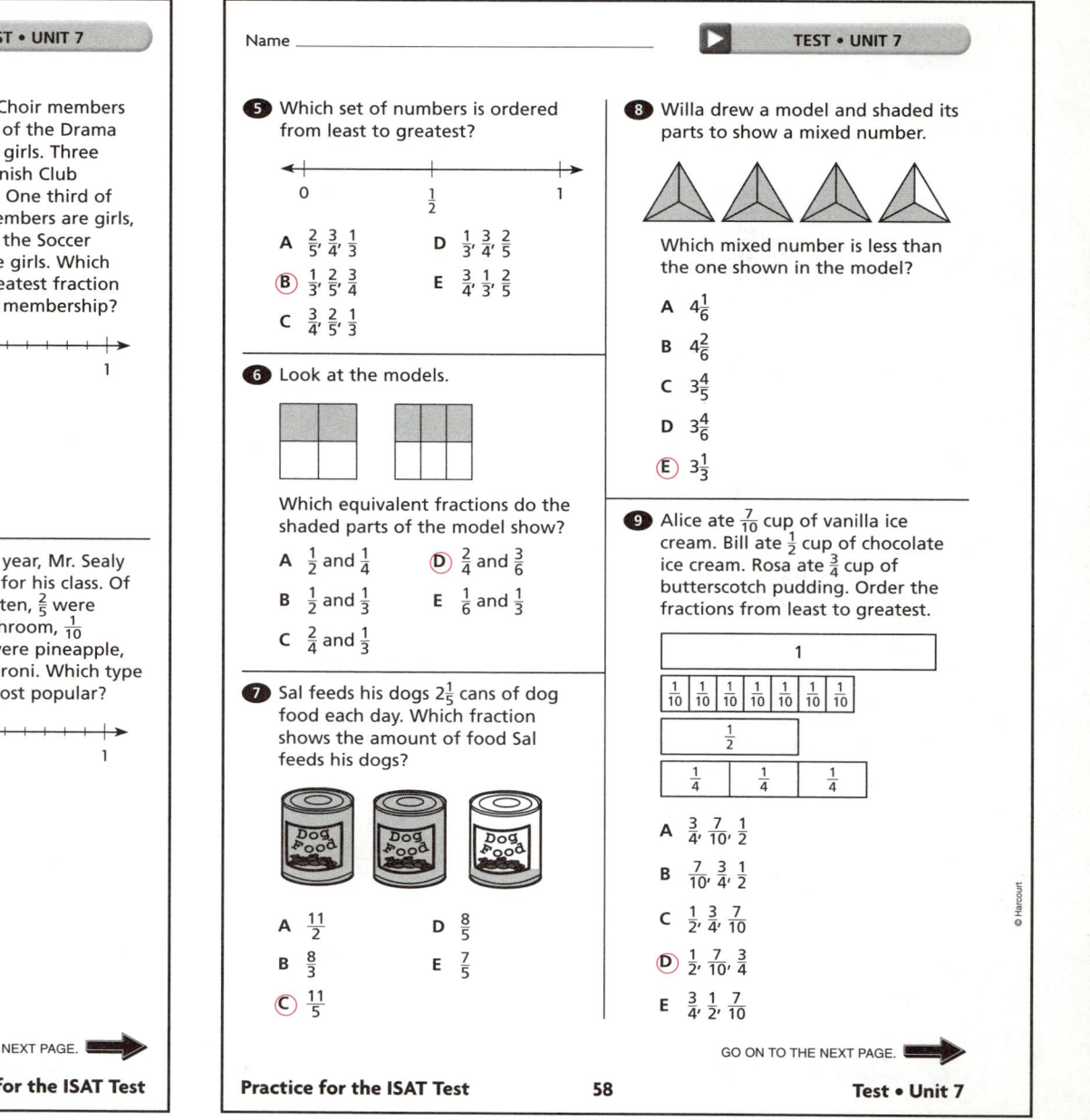

Name _____

1 A spinner has 16 equal sections. One fourth of the spinner is red, $\frac{3}{16}$ is blue, $\frac{3}{8}$ is green, $\frac{1}{16}$ is orange, and one eighth is yellow. Which color covers the greatest part of the spinner?

A blue
B green ⬤
C orange
D red
E yellow

2 Which number sentence does the model show?

A $\frac{4}{5} > \frac{8}{10}$
B $\frac{8}{10} < \frac{4}{5}$
C $\frac{4}{5} < \frac{8}{10}$
D $\frac{8}{10} > \frac{4}{5}$
E $\frac{4}{5} = \frac{8}{10}$ ⬤

3 Two thirds of the Choir members are girls. One half of the Drama Club members are girls. Three fourths of the Spanish Club members are girls. One third of the Track Team members are girls, and one fourth of the Soccer Team members are girls. Which activity has the greatest fraction representing girls' membership?

A Choir
B Drama Club
C Track Team
D Spanish Club ⬤
E Soccer Team

4 At the end of the year, Mr. Sealy had a pizza party for his class. Of the pizza slices eaten, $\frac{2}{5}$ were plain, $\frac{3}{8}$ were mushroom, $\frac{1}{10}$ were sausage, $\frac{1}{8}$ were pineapple, and $\frac{3}{5}$ were pepperoni. Which type of slice was the most popular?

A mushroom
B plain
C sausage
D pineapple
E pepperoni ⬤

5 Which set of numbers is ordered from least to greatest?

A $\frac{2}{5}, \frac{3}{5}, \frac{1}{3}$
B $\frac{1}{3}, \frac{2}{5}, \frac{3}{4}$ ⬤
C $\frac{3}{4}, \frac{2}{5}, \frac{1}{3}$
D $\frac{1}{3}, \frac{3}{4}, \frac{2}{5}$
E $\frac{3}{4}, \frac{1}{3}, \frac{2}{5}$

6 Look at the models.

Which equivalent fractions do the shaded parts of the model show?

A $\frac{1}{2}$ and $\frac{1}{4}$
B $\frac{1}{2}$ and $\frac{1}{3}$
C $\frac{2}{4}$ and $\frac{1}{3}$
D $\frac{2}{4}$ and $\frac{3}{6}$ ⬤
E $\frac{1}{6}$ and $\frac{1}{3}$

7 Sal feeds his dogs $2\frac{1}{5}$ cans of dog food each day. Which fraction shows the amount of food Sal feeds his dogs?

A $\frac{11}{2}$
B $\frac{8}{3}$
C $\frac{11}{5}$ ⬤
D $\frac{8}{5}$
E $\frac{7}{5}$

8 Willa drew a model and shaded its parts to show a mixed number.

Which mixed number is less than the one shown in the model?

A $4\frac{1}{6}$
B $4\frac{2}{6}$
C $3\frac{4}{5}$
D $3\frac{4}{6}$
E $3\frac{1}{3}$ ⬤

9 Alice ate $\frac{7}{10}$ cup of vanilla ice cream. Bill ate $\frac{1}{2}$ cup of chocolate ice cream. Rosa ate $\frac{3}{4}$ cup of butterscotch pudding. Order the fractions from least to greatest.

A $\frac{3}{4}, \frac{7}{10}, \frac{1}{2}$
B $\frac{7}{10}, \frac{3}{4}, \frac{1}{2}$
C $\frac{1}{2}, \frac{3}{4}, \frac{7}{10}$
D $\frac{1}{2}, \frac{7}{10}, \frac{3}{4}$ ⬤
E $\frac{3}{4}, \frac{1}{2}, \frac{7}{10}$

TEST • UNIT 7

10. Vanessa pitched $\frac{5}{9}$ of a softball game. Alicia pitched $\frac{2}{9}$ of the same game. How much more of the game did Vanessa pitch than Alicia?

A 3
B $\frac{7}{9}$
C $\frac{2}{3}$
D $\frac{1}{3}$ (circled)
E $\frac{1}{6}$

11. In a two-person relay race, the first runner will run $\frac{1}{8}$ of a mile. The second runner will then run $\frac{2}{8}$ of a mile. How far will they run in all?

A $\frac{3}{16}$ mile
B $\frac{3}{8}$ mile (circled)
C $\frac{1}{8}$ mile
D $\frac{1}{16}$ mile
E 1 mile

12. Four students have fraction cards. Kenny has the lowest fraction. Martha has the greatest fraction. Jack has a fraction equal to $\frac{1}{2}$. Vic has a fraction greater than Jack's. Which fraction could be Kenny's?

A $\frac{3}{7}$ (circled)
B $\frac{3}{6}$
C $\frac{3}{5}$
D $\frac{3}{4}$
E $\frac{9}{10}$

13. Mary ate $\frac{3}{4}$ of a pie. How much of the pie is left?

A $\frac{1}{4}$ of the pie (circled)
B $\frac{2}{5}$ of the pie
C $\frac{1}{2}$ of the pie
D $\frac{3}{5}$ of the pie
E $\frac{3}{3}$ of the pie

14. Mr. Garner is baking cookies. He put the first batch of 12 cookies on a rack to cool and went to get his mail. When he returned, only 9 cookies were left. What fraction of the cookies was missing?

A $\frac{3}{4}$ of the cookies
B $\frac{1}{2}$ of the cookies
C $\frac{1}{4}$ of the cookies (circled)
D $\frac{1}{12}$ of the cookies
E $\frac{1}{9}$ of the cookies

15. Which sum are you most likely to get if you spin this pointer 2 times?

A 11
B 10 (circled)
C 9
D 8
E 5

16. Will tossed two number cubes, each labeled from 1 to 6. How likely are his chances of rolling a sum of 1?

A certain
B very likely
C somewhat likely
D about the same as his chances of rolling a sum of 2
E impossible (circled)

17. Anson is going to take a marble from the bag without looking. What is the probability that he will choose a black marble?

A certain
B less likely than choosing a striped marble
C more likely than choosing a white marble
D equally likely as choosing a white marble (circled)
E unlikely

Name _____

18 The table shows how far each of 3 students walks to school every day.

WALKING DISTANCES

Student	Distance Walked
Greg	$\frac{3}{4}$ mile
Jason	$\frac{3}{8}$ mile
Sara	$\frac{1}{2}$ mile

On the lines below, explain how you can use a number line to find out who walks the greatest distance to school and who walks the least distance to school.

Possible explanation: I can graph all the fractions on a number line. The fraction farthest to the right shows the greatest distance. The fraction farthest to the left shows the least distance.

Show each distance on the number line. Label the points. Solve the problem.

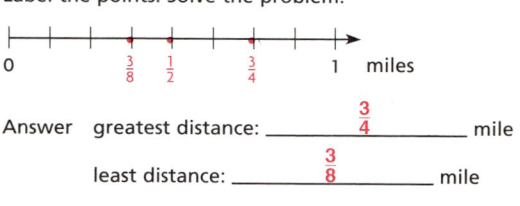

0   $\frac{3}{8}$ $\frac{1}{2}$ $\frac{3}{4}$ 1 miles

Answer   greatest distance: $\frac{3}{4}$ _____ mile

least distance: $\frac{3}{8}$ _____ mile

GO ON TO THE NEXT PAGE.

Name _____

19 Neena invited 5 friends to a pizza party. She decided to order 2 pizzas and divide each into 6 equal slices, so that each person could have 2 slices of pizza.

Draw the 2 pizzas, and shade parts to show the fraction of the pizzas that each person could have. Write a fraction for the shaded part.

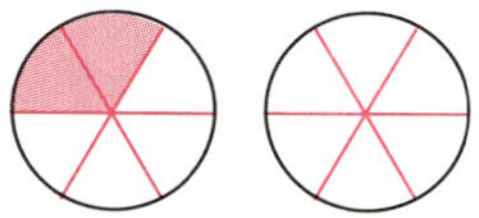

$\frac{2}{12}$, or $\frac{1}{6}$

On Thursday, 2 more friends said they would come. Neena decided to divide the 2 pizzas into 8 equal slices each. Draw the 2 pizzas, and shade parts to show the fraction of the pizzas that each person could have. Write a fraction for the shaded part.

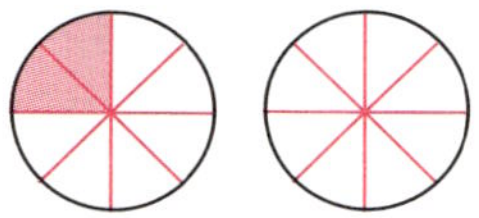

$\frac{2}{16}$, or $\frac{1}{8}$

On Friday, 4 more people said they would come, for a total of 12. Neena thought that she could divide the 2 pizzas into 12 equal slices each. That way each person could have $\frac{2}{12}$ of the pizzas. Describe her error. Write the correct answer.

There are 24 pieces of pizza, so each person can have $\frac{2}{24}$, or $\frac{1}{12}$, of the pizzas.

STOP

Name _____

TEST • UNIT 8

1 The table shows the voter turnout, expressed as a decimal, for each Presidential election from 1988 to 2000.

VOTER TURNOUT IN PRESIDENTIAL ELECTIONS

Year	Winner	Voter Turnout
1988	G.H.W. Bush	0.5
1992	W.J. Clinton	0.56
1996	W.J. Clinton	0.49
2000	G.W. Bush	0.51

Which year had the greatest voter turnout?

A 1988
(B) 1992
C 1996
D 2000
E Voter turnout was the same in 1988 and 1996.

2 Which of the following number sentences is true?

A 2.75 < 2.27
(B) 1.68 > 1.4
C 0.94 > 1.08
D 3.2 < 3.15
E 3.18 = 3.2

3 Which of the following number sentences is true?

0 ┼─┼─┼─┼─┼─┼─┼─┼─┼─┼─┼ 1.0
  0.1 0.2 0.3 0.4 0.5 0.6 0.7 0.8 0.9

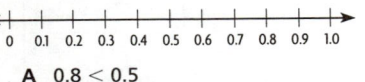

A 0.8 < 0.5
B 0.1 = 1.0
(C) 0.6 > 0.3
D 0.5 > 0.6
E 0.4 > 0.5

4 William wants to leave a tip that is more than 0.15 of his lunch check and less than 0.2 of his lunch check. Which decimal describes the tip he wants to leave? You may use the number line.

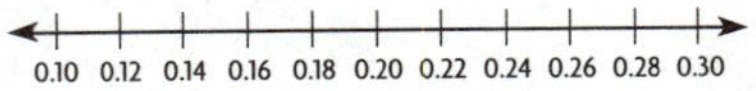

0.10 0.12 0.14 0.16 0.18 0.20 0.22 0.24 0.26 0.28 0.30

A 0.22
(B) 0.18
C 0.14
D 0.08
E 0.05

5 The large square in the model below stands for the number 1. The shaded portion of the model shows the length in decimeters of an Acteon beetle, which is the bulkiest insect in the world. Which decimal shows the length of an Acteon beetle?

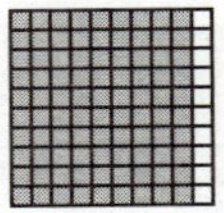

A 1.90 decimeters
B 0.99 decimeters
(C) 0.90 decimeters
D 0.10 decimeters
E 0.09 decimeters

Test • Unit 8   63   Practice for the ISAT Test

Name _____

TEST • UNIT 8

6 Katie used a hundredths model to show the total number of inches of rain reported in the last seven days.

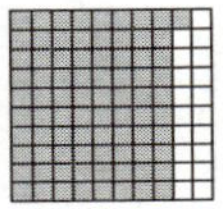

Which number sentence shows the amount of rain?

A 0.80 + 0.10 = 0.90
B 0.80 + 1.00 = 1.80
(C) 0.80 + 0.01 = 0.81
D 0.90 + 0.01 = 0.91
E 0.90 + 0.10 = 1.00

7 Look at the models.

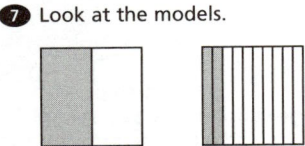

Which of the following is a true statement shown by the shaded parts of the models?

A $\frac{1}{2} = 0.2$
B $\frac{1}{2} < 0.2$
(C) $\frac{1}{2} > 0.2$
D $\frac{1}{2} = 0.02$
E $\frac{1}{2} > 0.02$

8 Which decimal has the same value as $\frac{1}{5}$?

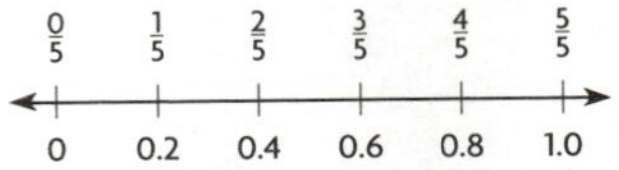

0 ─┼───┼───┼───┼───┼─ 1.0
   $\frac{1}{5}$ $\frac{2}{5}$ $\frac{3}{5}$ $\frac{4}{5}$

A 5.0
B 2.0
C 0.5
D 0.25
(E) 0.2

9 Melissa bought a skirt, a blouse, and a pair of shoes at the mall. How much did she spend in all?

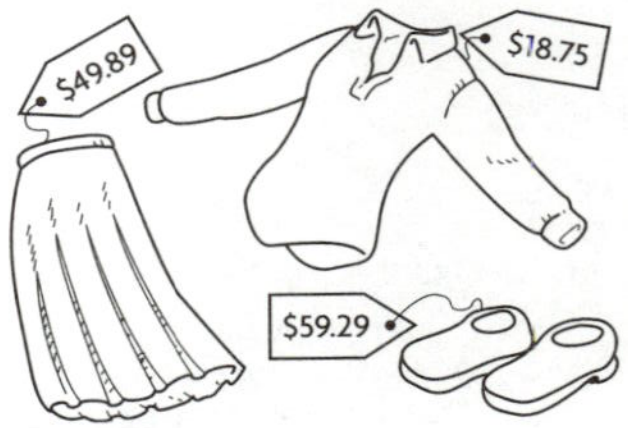

$18.75 $59.29 $49.89

A $106.73
B $117.83
C $126.83
D $126.93
(E) $127.93

10 Mr. Krutzler bought a tent for $99.87. The next week the tent was on sale for $88.46. How much higher was the price that Mr. Krutzler paid than the sale price?

A $17.41
B $12.40
(C) $11.41
D $11.34
E $10.41

Practice for the ISAT Test   64   Test • Unit 8

Name _____

TEST • UNIT 8

11 Marvin does not feel well, so his mother takes his temperature. The thermometer reads 101.2°F. A normal human temperature is 98.6°F. About how many degrees higher than normal is Marvin's temperature?

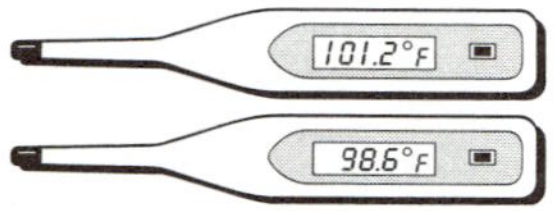

A about 1°F
(B) about 2°F
C about 4°F
D about 6°F
E about 8°F

12 The Kellys bought two sleeping bags. The Comfortplus cost $94.93 and the Overnighter cost $59.95. About how much more did they spend on the Comfortplus sleeping bag than on the Overnighter?

(A) $35
B $60
C $95
D $100
E $155

13 Miriam is weighing rocks for a science project. One rock weighs 17.29 pounds. Another rock weighs 0.49 pounds. About how much do both rocks weigh to the nearest tenth?

A 17 pounds
B 17.2 pounds
C 17.3 pounds
D 17.5 pounds
(E) 17.8 pounds

14 Use the model to help find the difference.

2.4 − 1.9

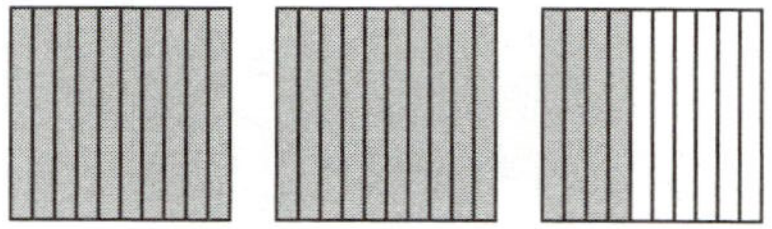

(A) 0.5
B 1.4
C 1.5
D 2.5
E 2.9

15 Which is the length of the paper clip?

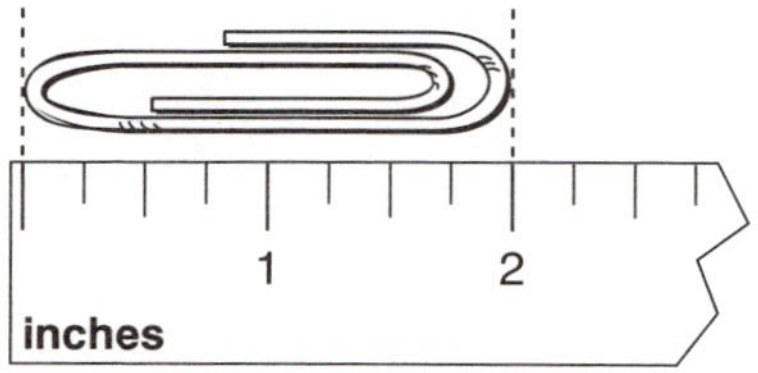

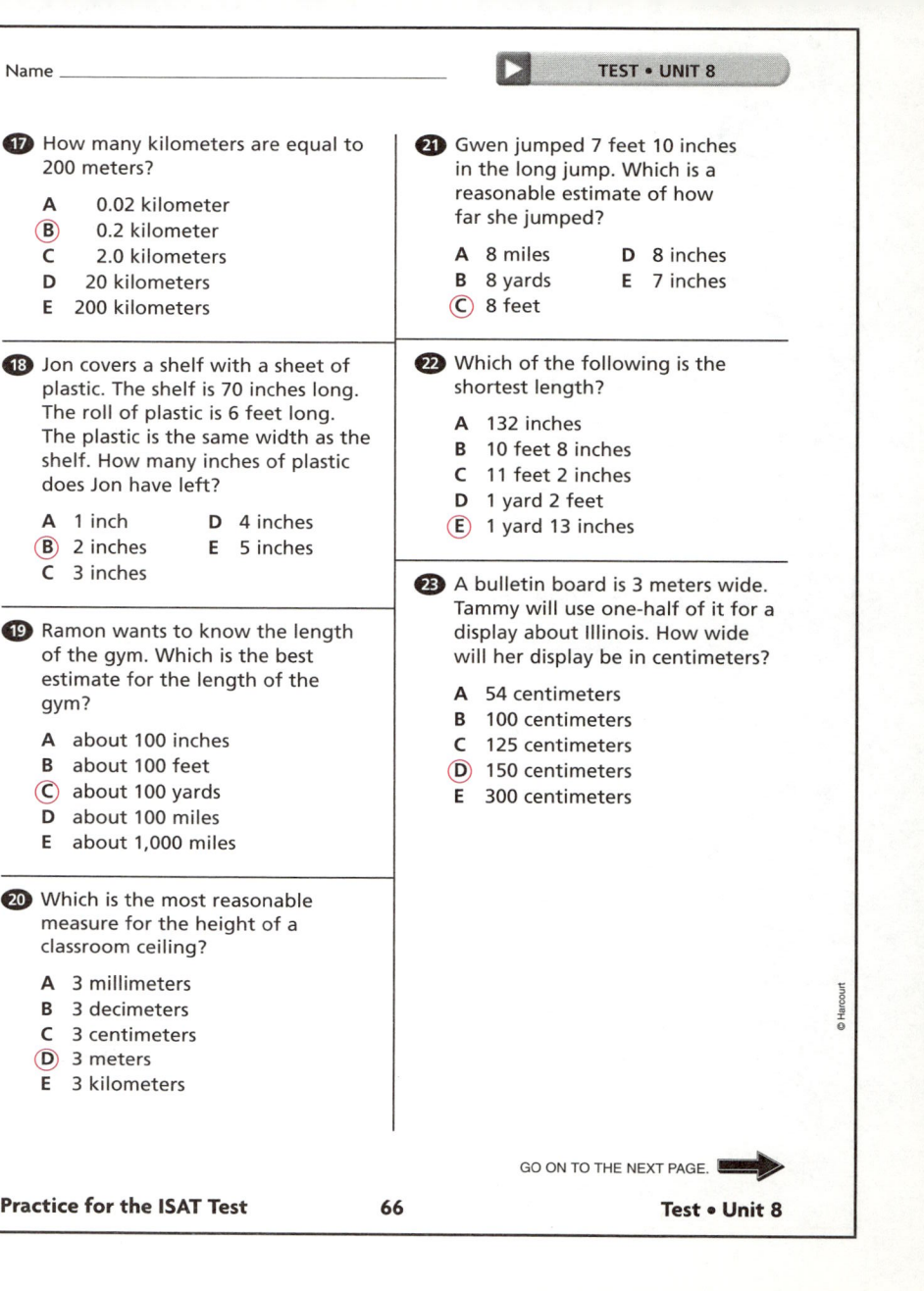

A $2\frac{1}{2}$ inches
(B) 2 inches
C $1\frac{1}{2}$ inches
D 1 inch
E $\frac{1}{2}$ inch

16 How many ounces are equal to 7 pounds?

A 56 ounces
B 84 ounces
C 105 ounces
(D) 112 ounces
E 142 ounces

GO ON TO THE NEXT PAGE.

Practice for the ISAT Test

Name _____

TEST • UNIT 8

17 How many kilometers are equal to 200 meters?

A 0.02 kilometer
(B) 0.2 kilometer
C 2.0 kilometers
D 20 kilometers
E 200 kilometers

18 Jon covers a shelf with a sheet of plastic. The shelf is 70 inches long. The roll of plastic is 6 feet long. The plastic is the same width as the shelf. How many inches of plastic does Jon have left?

A 1 inch
(B) 2 inches
C 3 inches
D 4 inches
E 5 inches

19 Ramon wants to know the length of the gym. Which is the best estimate for the length of the gym?

A about 100 inches
B about 100 feet
(C) about 100 yards
D about 100 miles
E about 1,000 miles

20 Which is the most reasonable measure for the height of a classroom ceiling?

A 3 millimeters
B 3 decimeters
C 3 centimeters
(D) 3 meters
E 3 kilometers

21 Gwen jumped 7 feet 10 inches in the long jump. Which is a reasonable estimate of how far she jumped?

A 8 miles
B 8 yards
(C) 8 feet
D 8 inches
E 7 inches

22 Which of the following is the shortest length?

A 132 inches
B 10 feet 8 inches
C 11 feet 2 inches
D 1 yard 2 feet
(E) 1 yard 13 inches

23 A bulletin board is 3 meters wide. Tammy will use one-half of it for a display about Illinois. How wide will her display be in centimeters?

A 54 centimeters
B 100 centimeters
C 125 centimeters
(D) 150 centimeters
E 300 centimeters

GO ON TO THE NEXT PAGE.

Practice for the ISAT Test

Name _____

TEST • UNIT 8

24 On a family vacation, Benito kept track of the mileage reported on the car's odometer. He took the first reading before the trip started.

| 1 | 0 | 6 | 8 | . | 9 |

At the end of the first day, the odometer read

| 1 | 4 | 1 | 6 | . | 3 |

How far did the family drive on the first day?

347.4 miles

At the end of the second day, the reading was

| 2 | 1 | 0 | 7 | . | 4 |

How far did the family drive on the second day?

691.1 miles

What was the total number of miles the family drove in 2 days?

1,038.5 miles

It is 900.5 miles to the next place that the family plans to visit on this vacation. Benito's family wants to take two days to drive this distance. If the family wants to drive more than 345.8 miles on the third day, how long should the third day's trip be? How many miles will the family need to drive on the fourth day?

Possible answer: 350 miles; 550.5 miles

GO ON TO THE NEXT PAGE.

Test • Unit 8   67   Practice for the ISAT Test

Name _____

TEST • UNIT 8

25 James and his dad are making pancakes for 25 campers. They need 1 cup of buttermilk for each batch of pancakes. How many batches can they make from 2 quarts of buttermilk?

Make a table to solve the problem.

Possible table:

| BUTTERMILK ||
Quarts	Pints	Cups
1	2	4
2	4	8

Answer: __**8**__ batches

Karen and Marty each have some juice to bring to the campout. The two have a total of 11 quarts of juice. Marty has 2 quarts more than twice the amount of juice that Karen has. How much juice do they each have?

Draw a diagram to solve.

Possible diagram:

Marty	3 quarts	3 quarts	2 quarts
Karen	3 quarts

Write your answer on the lines below.

Answer: Marty: __**8**__ quarts

Karen: __**3**__ quarts

Practice for the ISAT Test   68   Test • Unit 8

Name _____

TEST • UNIT 9

① Scott used 36 square yards of fertilizer to cover his backyard. The shape of his yard is square and its width is 6 yards. What is the length of the backyard?

- Ⓐ 6 yards
- B 9 yards
- C 12 yards
- D 18 yards
- E 24 yards

② The Murphys have a pond in their garden. If each square in the grid measures 1 square foot, what is the best estimate of the area of the pond?

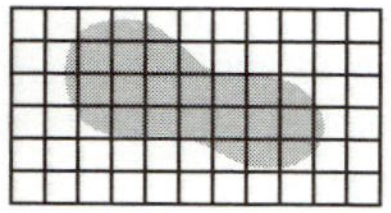

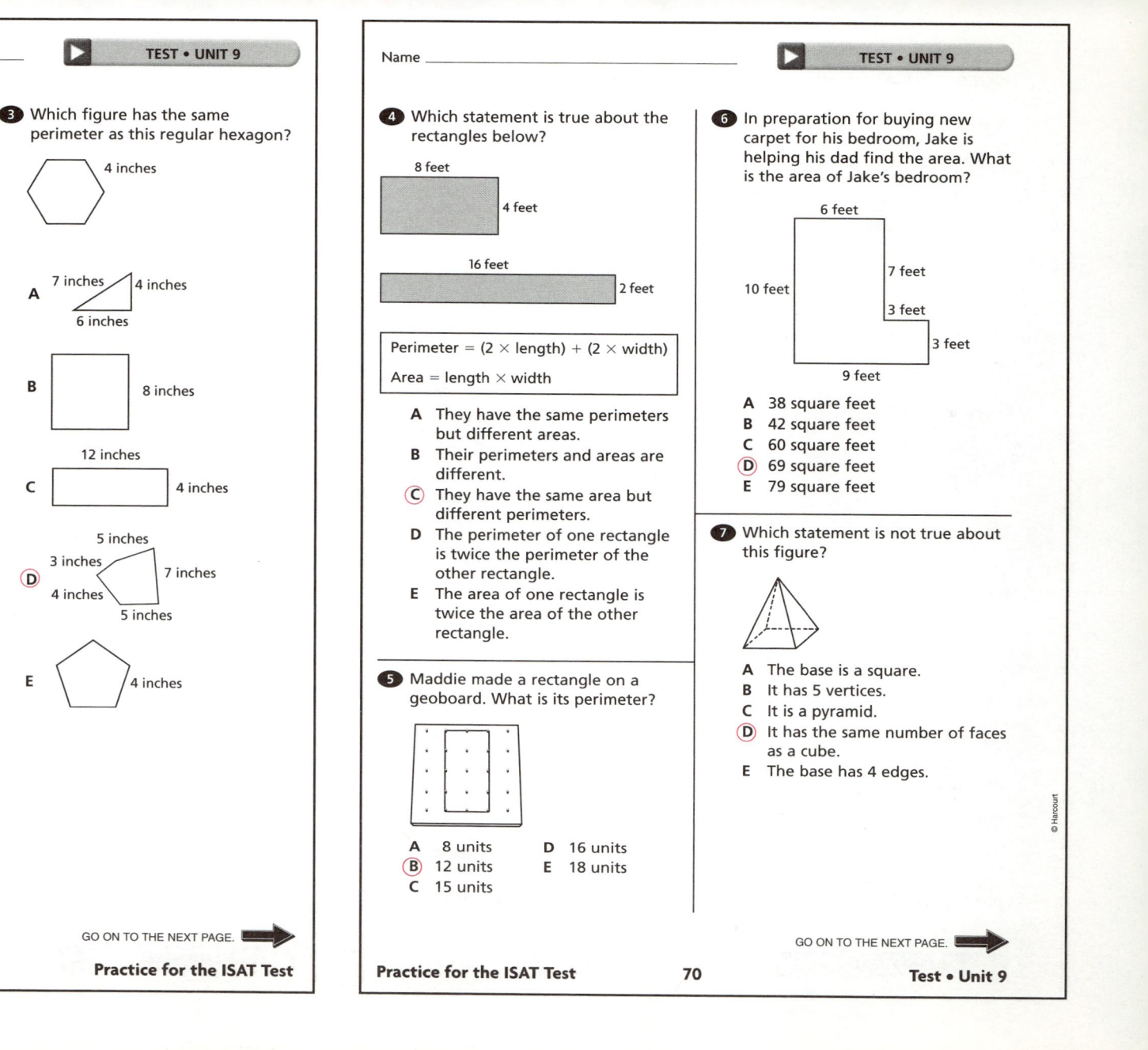

- Ⓐ $18\frac{1}{2}$ square feet
- B $27\frac{1}{2}$ square feet
- C 45 square feet
- D 66 square feet
- E 80 square feet

③ Which figure has the same perimeter as this regular hexagon?

4 inches

- A 7 inches / 4 inches / 6 inches (triangle)
- B 8 inches × 12 inches (rectangle)
- C 4 inches × 4 inches (rectangle)
- Ⓓ 3 inches, 4 inches, 5 inches, 5 inches, 7 inches (pentagon)
- E 4 inches (regular pentagon)

Test • Unit 9   69   Practice for the ISAT Test

Name _____

TEST • UNIT 9

④ Which statement is true about the rectangles below?

8 feet × 4 feet
16 feet × 2 feet

Perimeter = (2 × length) + (2 × width)
Area = length × width

- A They have the same perimeters but different areas.
- B Their perimeters and areas are different.
- Ⓒ They have the same area but different perimeters.
- D The perimeter of one rectangle is twice the perimeter of the other rectangle.
- E The area of one rectangle is twice the area of the other rectangle.

⑤ Maddie made a rectangle on a geoboard. What is its perimeter?

- A 8 units
- Ⓑ 12 units
- C 15 units
- D 16 units
- E 18 units

⑥ In preparation for buying new carpet for his bedroom, Jake is helping his dad find the area. What is the area of Jake's bedroom?

- A 38 square feet
- B 42 square feet
- C 60 square feet
- Ⓓ 69 square feet
- E 79 square feet

⑦ Which statement is not true about this figure?

- A The base is a square.
- B It has 5 vertices.
- C It is a pyramid.
- Ⓓ It has the same number of faces as a cube.
- E The base has 4 edges.

Practice for the ISAT Test   70   Test • Unit 9

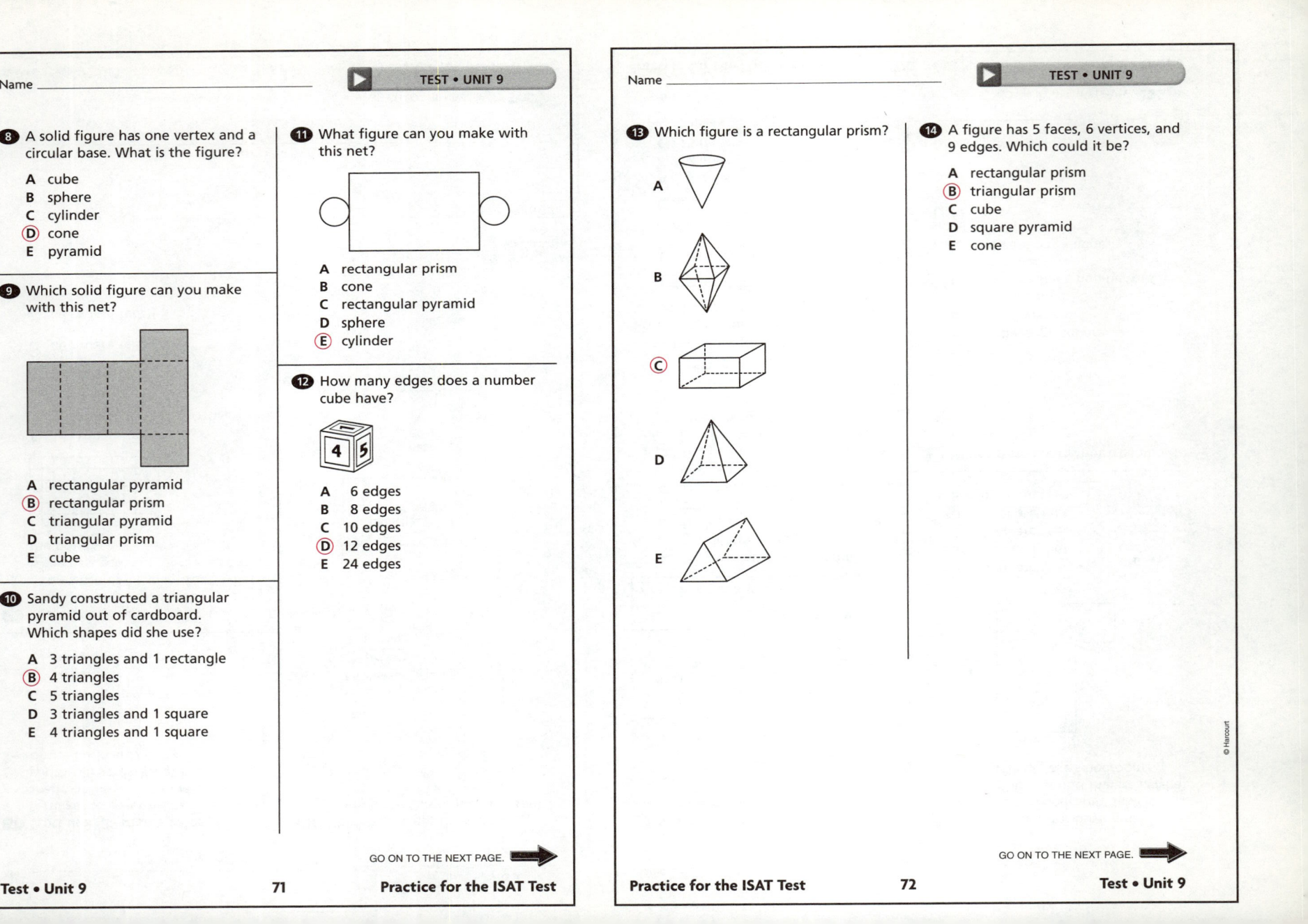

Name _____

8 A solid figure has one vertex and a circular base. What is the figure?

A cube
B sphere
C cylinder
(D) cone
E pyramid

9 Which solid figure can you make with this net?

A rectangular pyramid
(B) rectangular prism
C triangular pyramid
D triangular prism
E cube

10 Sandy constructed a triangular pyramid out of cardboard. Which shapes did she use?

A 3 triangles and 1 rectangle
(B) 4 triangles
C 5 triangles
D 3 triangles and 1 square
E 4 triangles and 1 square

11 What figure can you make with this net?

A rectangular prism
B cone
C rectangular pyramid
D sphere
(E) cylinder

12 How many edges does a number cube have?

A 6 edges
B 8 edges
C 10 edges
(D) 12 edges
E 24 edges

Test • Unit 9   71   Practice for the ISAT Test

GO ON TO THE NEXT PAGE.

Name _____

13 Which figure is a rectangular prism?

A
B
(C)
D
E

14 A figure has 5 faces, 6 vertices, and 9 edges. Which could it be?

A rectangular prism
(B) triangular prism
C cube
D square pyramid
E cone

Practice for the ISAT Test   72   Test • Unit 9

GO ON TO THE NEXT PAGE.

Name _____

15 Jana's watercolor paints come in a square container.

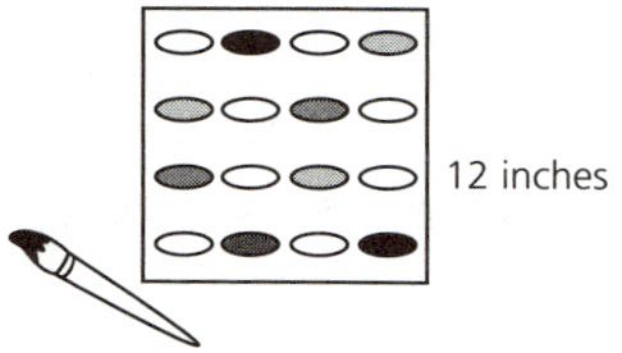

12 inches

Draw a figure that has the same perimeter as the base of the container but a different area. Check students' work.
Possible answer:

4 inches
20 inches

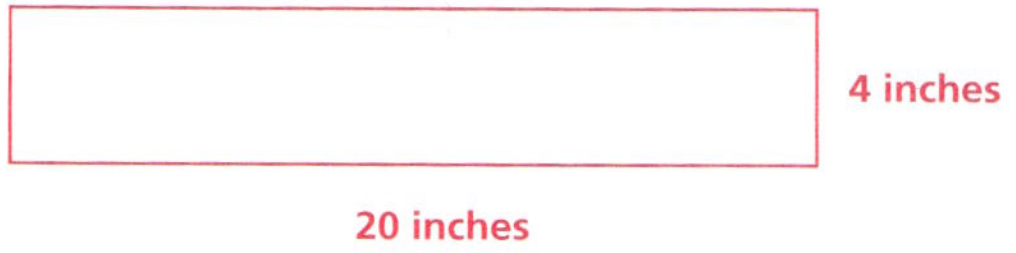

The perimeter is 48 inches, and its area is 80 square inches.

Draw a figure that has the same area as the base of the container but a different perimeter. Check students' work.
Possible answer:

8 inches
18 inches

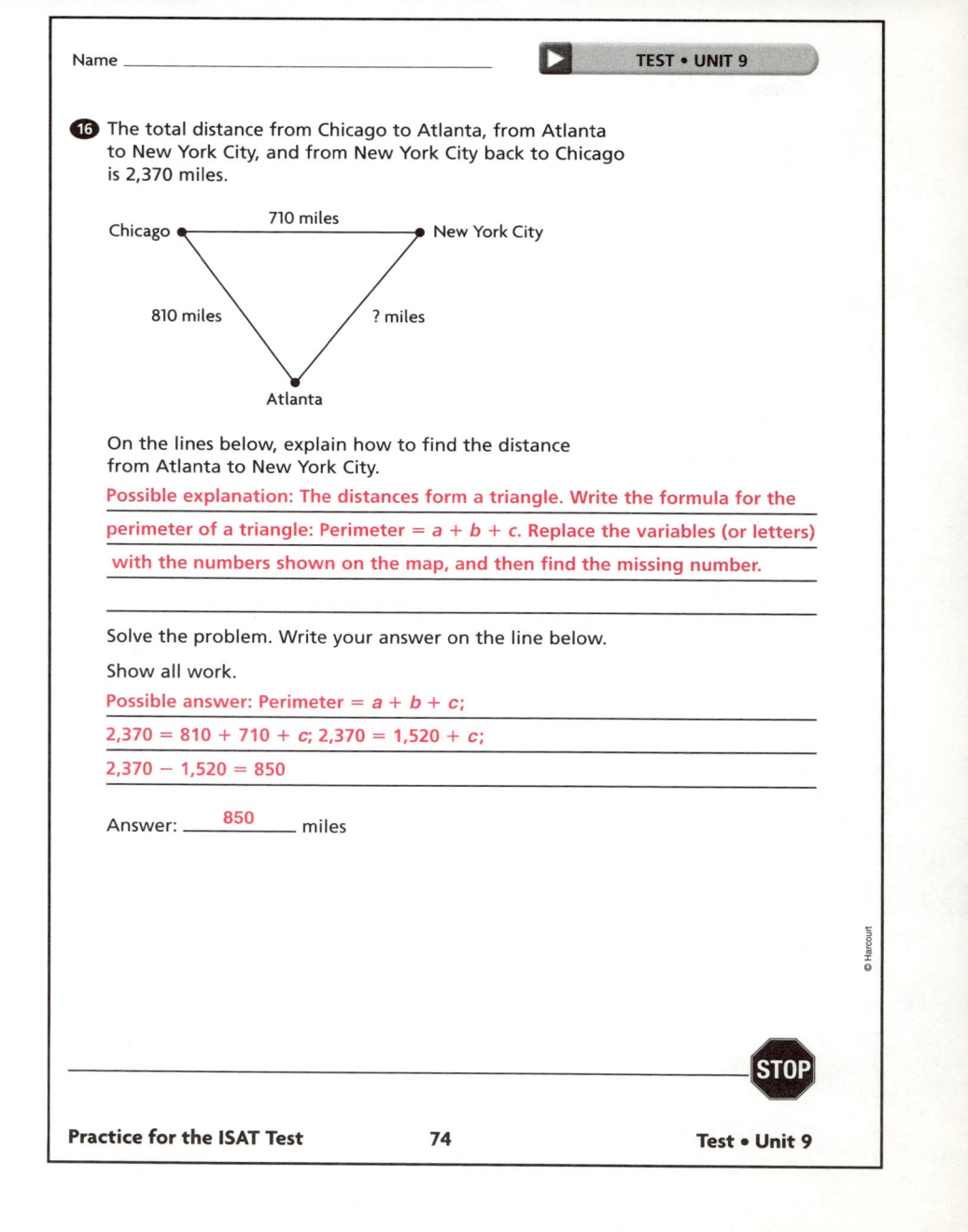

The perimeter is 52 inches, and its area is 144 square inches.

On the lines below, explain how you found your answers.
Possible answer: I figured out the perimeter and area of the square base. The perimeter is 4 × 12 = 48 inches. The area is 12 × 12 = 144 square inches. For the first rectangle, I divided 48 by 2 to get 24, and made the length plus width equal to 24. For the area, I found 2 factors of 48 that I could use for the length and width of the second rectangle.

Test • Unit 9   73   Practice for the ISAT Test

GO ON TO THE NEXT PAGE.

Name _____

16 The total distance from Chicago to Atlanta, from Atlanta to New York City, and from New York City back to Chicago is 2,370 miles.

Chicago
810 miles
710 miles
New York City
? miles
Atlanta

On the lines below, explain how to find the distance from Atlanta to New York City.
Possible explanation: The distances form a triangle. Write the formula for the perimeter of a triangle: Perimeter = $a + b + c$. Replace the variables (or letters) with the numbers shown on the map, and then find the missing number.

Solve the problem. Write your answer on the line below. Show all work.
Possible answer: Perimeter = $a + b + c$;
$2,370 = 810 + 710 + c$; $2,370 = 1,520 + c$;
$2,370 − 1,520 = 850$

Answer: ___850___ miles

Practice for the ISAT Test   74   Test • Unit 9

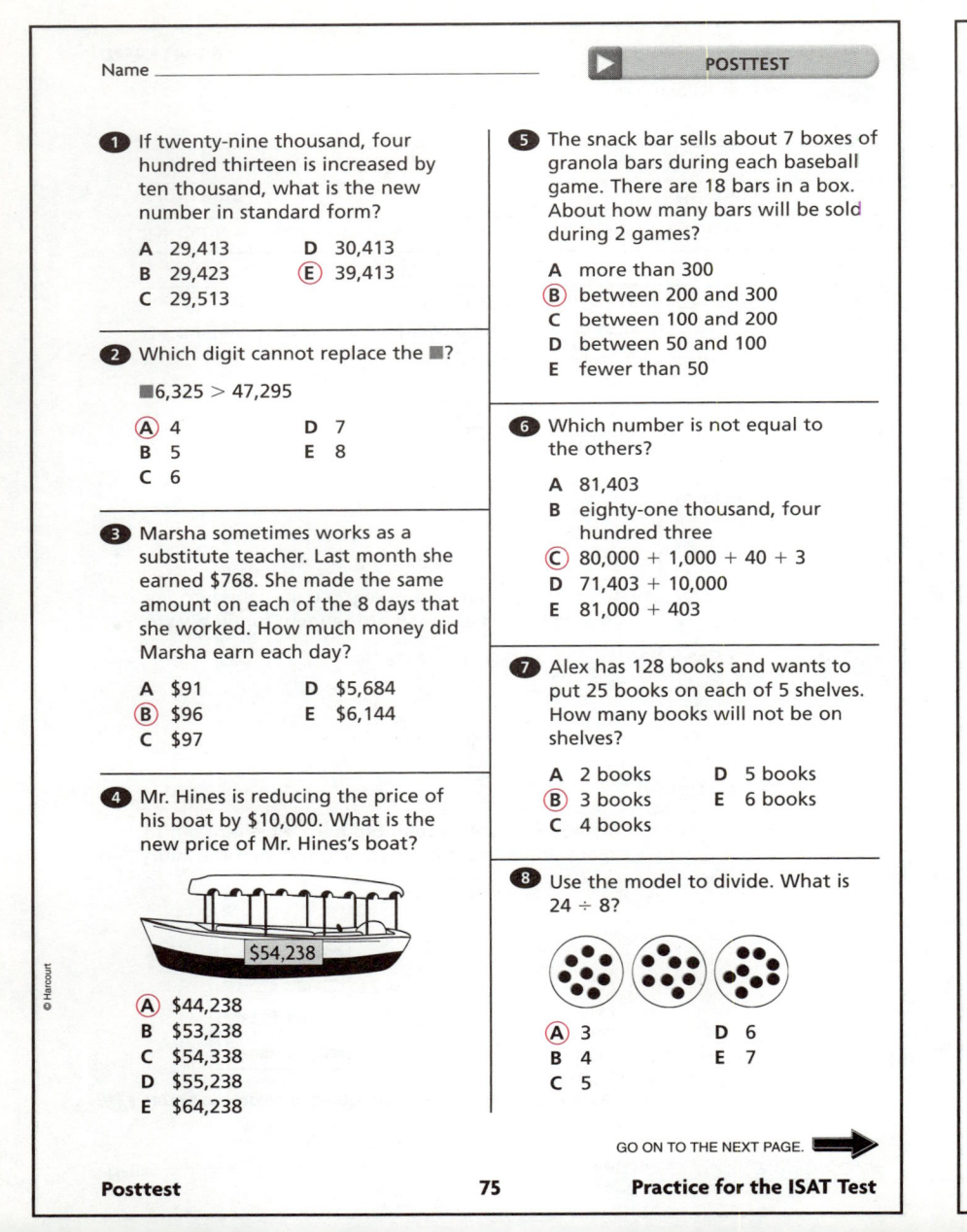

Posttest

1. If twenty-nine thousand, four hundred thirteen is increased by ten thousand, what is the new number in standard form?

   A 29,413
   B 29,423
   C 29,513
   D 30,413
   E 39,413

2. Which digit cannot replace the ■?

   ■6,325 > 47,295

   A 4
   B 5
   C 6
   D 7
   E 8

3. Marsha sometimes works as a substitute teacher. Last month she earned $768. She made the same amount on each of the 8 days that she worked. How much money did Marsha earn each day?

   A $91
   B $96
   C $97
   D $5,684
   E $6,144

4. Mr. Hines is reducing the price of his boat by $10,000. What is the new price of Mr. Hines's boat?

   $54,238

   A $44,238
   B $53,238
   C $54,338
   D $55,238
   E $64,238

5. The snack bar sells about 7 boxes of granola bars during each baseball game. There are 18 bars in a box. About how many bars will be sold during 2 games?

   A more than 300
   B between 200 and 300
   C between 100 and 200
   D between 50 and 100
   E fewer than 50

6. Which number is not equal to the others?

   A 81,403
   B eighty-one thousand, four hundred three
   C 80,000 + 1,000 + 40 + 3
   D 71,403 + 10,000
   E 81,000 + 403

7. Alex has 128 books and wants to put 25 books on each of 5 shelves. How many books will not be on shelves?

   A 2 books
   B 3 books
   C 4 books
   D 5 books
   E 6 books

8. Use the model to divide. What is $24 \div 8$?

   A 3
   B 4
   C 5
   D 6
   E 7

Practice for the ISAT Test — 75

9. Which division equation is shown by the model?

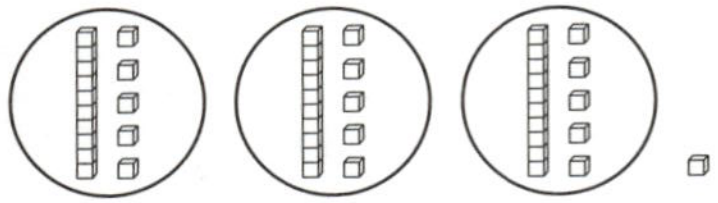

   A $46 \div 3 = 14$
   B $46 \div 3 = 14$ r4
   C $42 \div 3 = 14$
   D $46 \div 3 = 15$ r1
   E $44 \div 3 = 14$ r4

10. A book has 267 pages. If each section is 9 pages, about how many sections are in the book?

    A about 20 sections
    B about 24 sections
    C about 30 sections
    D about 35 sections
    E about 39 sections

11. Gerald had $\frac{1}{2}$ gallon of milk. He used $\frac{3}{8}$ of the milk to make pancakes.

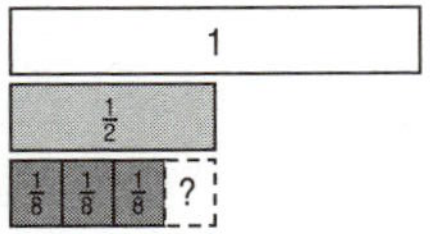

    Use the fraction bars to find out how much milk is left.

    A $\frac{1}{8}$ gallon
    B $\frac{1}{2}$ gallon
    C $\frac{1}{4}$ gallon
    D $\frac{3}{8}$ gallon
    E $\frac{1}{10}$ gallon

12. The shaded parts of this model show a mixed number. Which mixed number is greater than the one in the model?

    A $1\frac{1}{3}$
    B $2\frac{1}{8}$
    C $2\frac{1}{4}$
    D $2\frac{1}{5}$
    E $2\frac{3}{5}$

13. Which number completes the pattern?

    $56 \div 8 = 7$
    $560 \div 8 = 70$
    $5,600 \div 8 = 700$
    $56,000 \div 8 = $ ■

    A 70
    B 77
    C 700
    D 750
    E 7,000

14. Mark has a card collection. Baseball cards make up $\frac{3}{8}$ of his collection. Football cards make up $\frac{1}{4}$ of his collection. Which fraction of Mark's collection is either baseball or football cards?

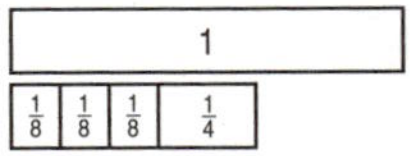

    A $\frac{1}{4}$
    B $\frac{3}{8}$
    C $\frac{1}{2}$
    D $\frac{5}{8}$
    E $\frac{3}{4}$

Practice for the ISAT Test — 76

POSTTEST

15 Which fraction is equivalent to the shaded part of the square?

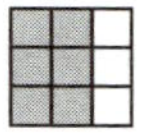

A $\frac{1}{3}$
B $\frac{1}{2}$
(C) $\frac{2}{3}$
D $\frac{3}{4}$
E $\frac{8}{9}$

16 Horatio bought seventy-five hundredths pound of nuts to use for his apple tarts for the school fair.

What is the number of pounds of nuts he bought in standard form?

A 0.075
(B) 0.75
C 7.5
D 75
E 750

17 Rosemary made 8 of her 10 free throw attempts in the last basketball game. Which decimal represents the number of free throws she made?

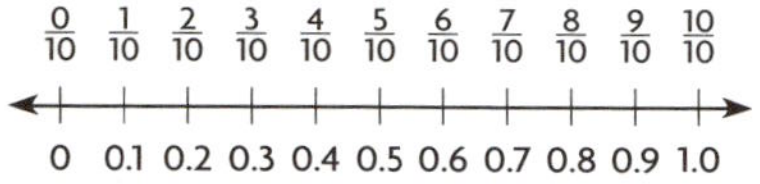

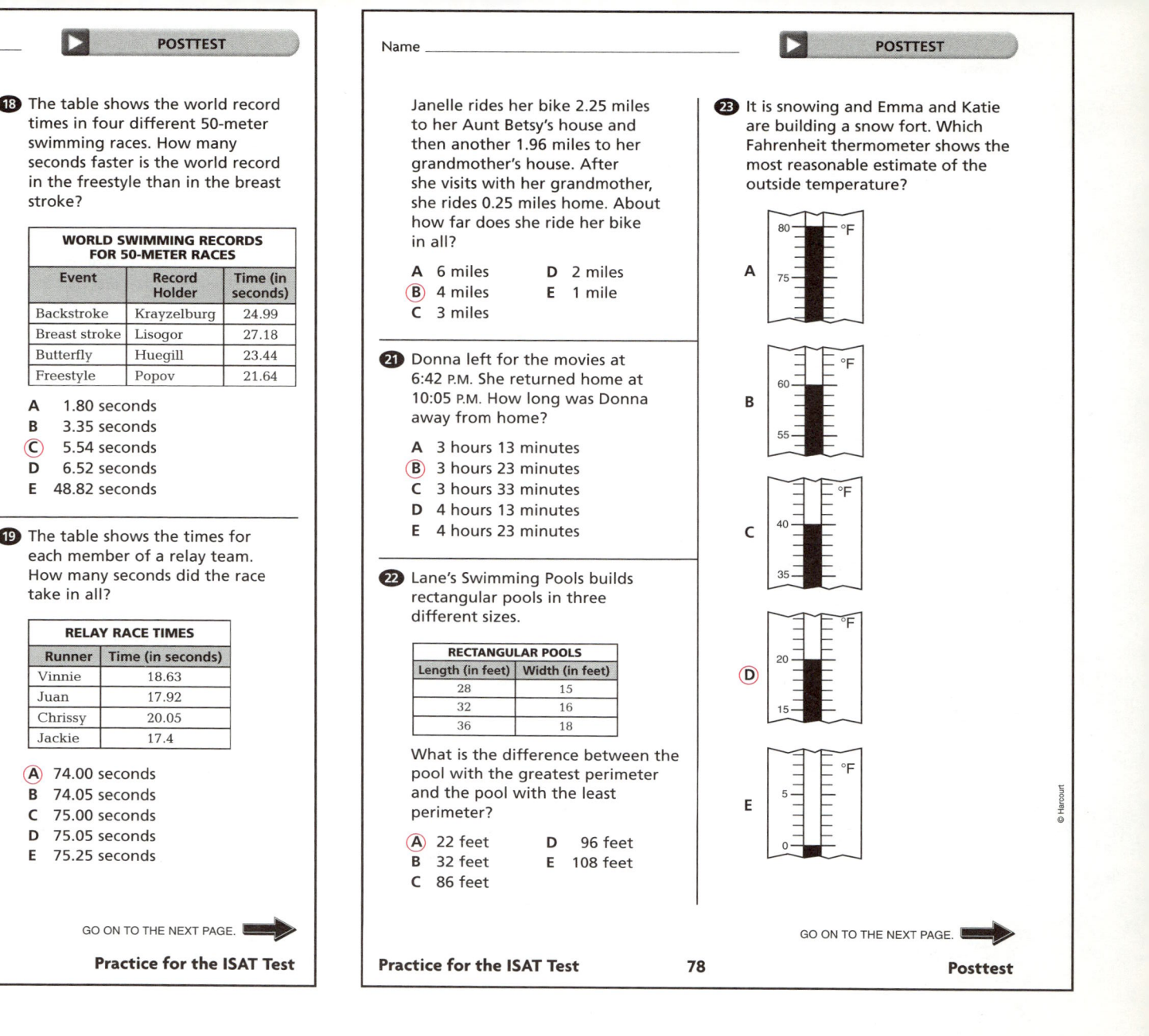

(A) 0.8
B 0.2
C 0.08
D 0.02
E 8.10

18 The table shows the world record times in four different 50-meter swimming races. How many seconds faster is the world record in the freestyle than in the breast stroke?

WORLD SWIMMING RECORDS FOR 50-METER RACES

Event	Record Holder	Time (in seconds)
Backstroke	Krayzelburg	24.99
Breast stroke	Lisogor	27.18
Butterfly	Huegill	23.44
Freestyle	Popov	21.64

A 1.80 seconds
B 3.35 seconds
(C) 5.54 seconds
D 6.52 seconds
E 48.82 seconds

19 The table shows the times for each member of a relay team. How many seconds did the race take in all?

RELAY RACE TIMES

Runner	Time (in seconds)
Vinnie	18.63
Juan	17.92
Chrissy	20.05
Jackie	17.4

(A) 74.00 seconds
B 74.05 seconds
C 75.00 seconds
D 75.05 seconds
E 75.25 seconds

---

20 Janelle rides her bike 2.25 miles to her Aunt Betsy's house and then another 1.96 miles to her grandmother's house. After she visits with her grandmother, she rides 0.25 miles home. About how far does she ride her bike in all?

A 6 miles
(B) 4 miles
C 3 miles
D 2 miles
E 1 mile

21 Donna left for the movies at 6:42 P.M. She returned home at 10:05 P.M. How long was Donna away from home?

A 3 hours 13 minutes
(B) 3 hours 23 minutes
C 3 hours 33 minutes
D 4 hours 13 minutes
E 4 hours 23 minutes

22 Lane's Swimming Pools builds rectangular pools in three different sizes.

RECTANGULAR POOLS

Length (in feet)	Width (in feet)
28	15
32	16
36	18

What is the difference between the pool with the greatest perimeter and the pool with the least perimeter?

(A) 22 feet
B 32 feet
C 86 feet
D 96 feet
E 108 feet

23 It is snowing and Emma and Katie are building a snow fort. Which Fahrenheit thermometer shows the most reasonable estimate of the outside temperature?

A (80–75 °F)
B (60–55 °F)
C (40–35 °F)
(D) (20–15 °F)
E (5–0 °F)

POSTTEST

Name _____

24 The graph shows the amount of money raised by the Art Club for its annual fund-raiser. How much more money did Tina raise than Chloe?

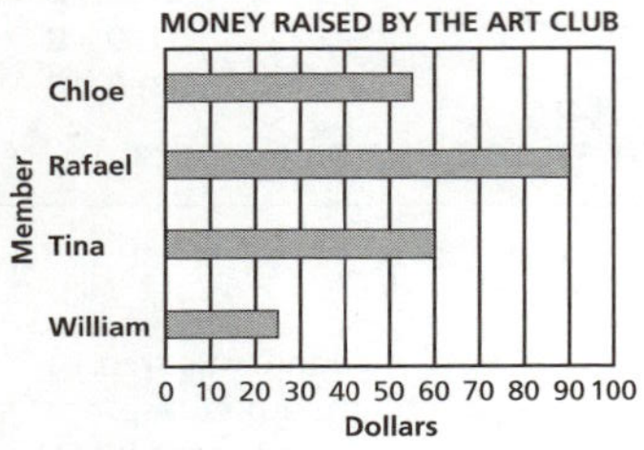

MONEY RAISED BY THE ART CLUB

A $115
B $35
C $30
D $10
E $5

25 Maria is using a 2-cup measuring cup to measure enough water to make 2 gallons of lemonade. How many times will she have to fill the measuring cup?

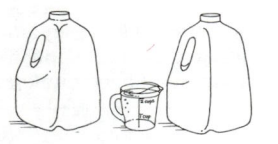

A 32 times
B 24 times
C 16 times
D 12 times
E 8 times

26 Sushi is a Japanese food. Fish or vegetables are wrapped inside a roll made of rice and seaweed called nori. A piece of nori weighs about 25 grams. Which is the best estimate for the weight of 40 pieces of nori?

A about 1,000 kilograms
B about 100 kilograms
C about 10 kilograms
D about 1,000 grams
E about 100 grams

27 This table shows how many feet are in a yard. What are the missing numbers in the table?

Yards	2	4	6	8	10	12
Feet	6	12	18	■	■	■

A 18, 21, 24   D 32, 36, 40
B 24, 30, 36   E 32, 40, 48
C 30, 36, 42

28 Use mental math. Which value for the variable makes the equation true?

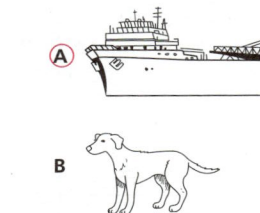

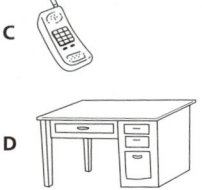

$16 + p = 23$

A $p = 40$ **D** $p = 7$
B $p = 39$ E $p = 6$
C $p = 9$

GO ON TO THE NEXT PAGE.

Posttest   79   Practice for the ISAT Test

POSTTEST

Name _____

29 Which object would be best measured in tons?

A (ship)
B (dog)
C (cell phone)
D (desk)
E (car)

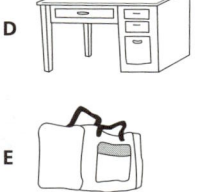

30 What is the area of the irregular polygon.

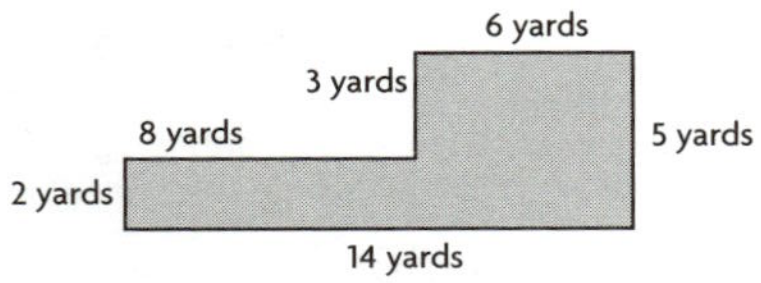

(8 yards, 2 yards, 3 yards, 6 yards, 5 yards, 14 yards)

A 24 square yards
B 30 square yards
C 38 square yards
D 46 square yards
E 84 square yards

31 Find a rule for the pattern in the input/output table.

Input a	Output b
7	23
15	31
28	44
43	59

A $b + 16 = a$ D $a + b = 16$
B $a + 16 = b$ E $16 - b = a$
C $a - 16 = b$

32 If the spinner is spun, how likely is it that the pointer will stop on a prime number?

A certain
B likely
C equally likely
D unlikely
E impossible

33 In the summer, a grizzly bear can gain 30 pounds of weight in one week to prepare for hibernation. If the weight gain is the same each week, how much weight will the bear have gained after 4 weeks?

GRIZZLY BEAR WEIGHT GAIN

Week	1	2	3	4
Pounds	30	60	90	■

A 100 pounds
B 110 pounds
C 120 pounds
D 130 pounds
E 140 pounds

GO ON TO THE NEXT PAGE.

Practice for the ISAT Test   80   Posttest

Posttest

34 Which is the missing number in the input/output table?

Input	Output
2	8
5	20
7	28
9	■

A 45
(B) 36
C 32
D 28
E 15

35 Rob saves some of his allowance each week. He has saved $22. Then, he decides to spend some of his money to see a movie. Then, he gets his weekly allowance of $10. The variable b stands for how much he spends for the movie. Which expression represents the situation?

A $(b + 22) - 10$
B $(22 + b) - 10$
(C) $(22 - b) + 10$
D $(b - 22) + 10$
E $(22 - 10) + b$

36 What is the missing number?

$(7 \times 6) \times 5 = 7 \times (\blacksquare \times 5)$

A 5
(B) 6
C 7
D 35
E 210

37 Which figure is not a regular polygon?

A (pentagon)
B (square)
C (hexagon)
D (triangle)
(E) (rectangle)

38 Bonnie built a fence in the shape of this figure. Which best describes the figure?

A 4 equal sides
(B) opposite sides are parallel
C opposite sides are perpendicular
D 3 equal sides
E perpendicular sides are equal

---

39 Quinn walks 5 blocks east. Then she walks 7 blocks south, 2 blocks west, 3 blocks north, and 5 blocks east. How many blocks does Quinn walk before she crosses her own path?

A 22 blocks
(B) 19 blocks
C 18 blocks
D 17 blocks
E 16 blocks

40 What building is located at (7,4)?

A bank
B hospital
(C) post office
D store
E school

41 This line graph shows the life expectancy of people living in the U.S. from 1910 to 2000. Which question cannot be answered by reading the graph?

A What was the life expectancy in 1940?
B In which year was the life expectancy 70?
C Has the life expectancy increased or decreased from 1910 to 2000?
D How much longer was the life expectancy in 2000 than in 1910?
(E) How much greater was the birth rate in 2000 than in 1940?

42 This table shows how many inches a cornstalk grows each day. Which is the mode of the data?

Daily Growth (in inches)

Sun	Mon	Tues	Wed	Thu	Fri	Sat
2	3	2	2	6	3	4

(A) 2 inches
B 3 inches
C 4 inches
D 5 inches
E 6 inches

POSTTEST

Name _____

43 A toy store has these items in the window.

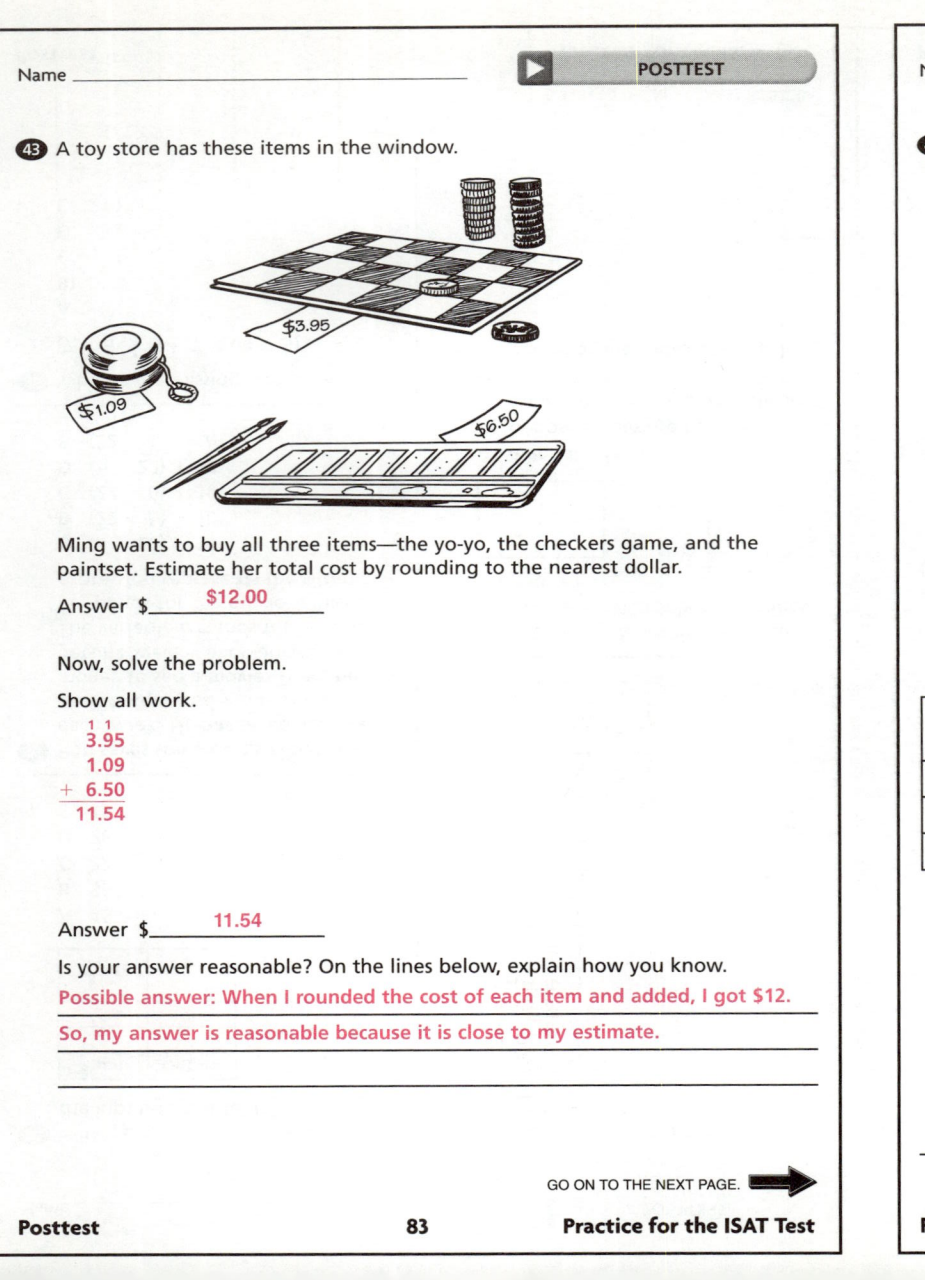

$3.95
$6.50
$1.09

Ming wants to buy all three items—the yo-yo, the checkers game, and the paintset. Estimate her total cost by rounding to the nearest dollar.

Answer $ __$12.00__

Now, solve the problem.

Show all work.
```
  1 1
  3.95
  1.09
+ 6.50
------
 11.54
```

Answer $ __11.54__

Is your answer reasonable? On the lines below, explain how you know.

Possible answer: When I rounded the cost of each item and added, I got $12.
So, my answer is reasonable because it is close to my estimate.

GO ON TO THE NEXT PAGE.

POSTTEST

Name _____

44 A candy manufacturer wants to arrange chocolates in rectangular boxes. The base of each box has an area of 24 square inches.

This grid is made up of square units. Draw three different rectangles that have an area of 24 square units. Number each rectangle.

Possible answer: Possible rectangles are 1 unit by 24 units, 2 units by 12 units, 3 units by 8 units, and 4 units by 6 units.

Make a table to show the width, length, perimeter, and area of each rectangle you drew. Possible answer:

Rectangle	Width (in units)	Length (in units)	Perimeter (in units)	Area (in square units)
1	2	12	28	24
2	4	6	20	24
3	3	8	22	24

The manufacturer adds a colored strip of paper around the top edge of the box. For which box would the manufacturer need the least amount of paper? Explain how you know.

Possible answer: The manufacturer would need the least amount for rectangle 2 because it has the smallest perimeter.

STOP

Correlation to the Illinois Mathematics Assessment Framework—Grade 4

MATHEMATICS ASSESSMENT FRAMEWORK

Objective Number	Objective	Test Item Location	Test Item Number(s)
Goal 6: Number Sense			
6.4.01 (A)	Compare the numerical value of two fractions having like and unlike denominators up to twelfths, using concrete or pictorial models involving areas/regions, lengths/measurements, and sets.	Unit 7	1, 2
6.4.02 (A)	Order and compare whole numbers and decimals to two decimal places.	Unit 1 Unit 8	1, 2 1, 2
6.4.03 (A)	Interpret whole numbers up to 100,000; demonstrate an understanding of the values of the digits and comparing and ordering the numbers.	Unit 1 Pretest Posttest	3, 4 1, 4 1, 4
6.4.04 (A)	Represent, order, and compare large numbers (up to 100,000) using various forms, including expanded notation (e.g., 853 = 8 × 100 + 5 × 10 + 3).	Unit 1 Pretest Posttest	5, 25 2 2
6.4.05 (A)	Identify on a number line the relative position of positive fractions, positive mixed numbers, and positive decimals to two decimal places.	Unit 7 Unit 8	3, 4, 18 3, 4
6.4.06 (A)	Exhibit an understanding of the base-ten number system by reading, naming, and writing decimals between 0 and 1 up through the hundredths.	Unit 8 Pretest Posttest	5 16 16
6.4.07 (A)	Perform prime factorization of all whole numbers through 20.	Unit 5	1, 3, 4
6.4.08 (A)	Identify all prime numbers through 20.	Unit 5	2, 5, 6
6.4.09 (A)	Identify classes (i.e., odds/evens, factors/multiples, squares) to which a number may belong, and identify the numbers in those classes. Use these in the solution of problems.	Unit 5	7, 8, 10
6.4.10 (A)	Recognize equivalent representations for decimals and generate them by composing and decomposing numbers (e.g., 0.15 = 0.1 + 0.05).	Unit 8	6
6.4.11 (D)	Select, use, and explain models to relate common fractions and mixed numbers (in halves, thirds, fourths, fifths, sixths, eighths and tenths); find equivalent fractions, mixed numbers, improper fractions, and decimals; and order fractions.	Unit 7 Unit 8 Pretest Posttest	5, 6, 7, 8, 9 7 11, 12 11, 12
6.4.12 (D)	Identify and generate equivalent forms of common decimals and fractions less than one whole (halves, quarters, fifths, and tenths).	Unit 8 Pretest Posttest	8 14, 17 14, 17
6.4.13 (B/C)	Add and subtract fractions with like denominators in simple computations or in word problems.	Unit 7	10, 11
6.4.14 (B/C)	Add and subtract decimals through hundredths.	Unit 8 Pretest Posttest	9, 10, 24 18, 19, 43 18, 19, 43

Standards Correlation   **Practice for the ISAT Test**

MATHEMATICS ASSESSMENT FRAMEWORK

Objective Number	Objective	Test Item Location	Test Item Number(s)
6.4.15 (B/C)	Make estimates appropriate to a given situation with whole numbers, fractions, and decimals by knowing when to estimate, and select the appropriate type of estimate including overestimate, underestimate, and range of estimate, and select the appropriate method of estimation.	Unit 1 Unit 4 Unit 5 Unit 8 Pretest Posttest	7, 8, 12 1, 2, 3, 4, 5 9, 11, 12, 14 11, 12, 13 5, 10, 20 5, 10, 20
6.4.16 (B/C)	Compute with whole numbers: addition—up to three 3-digit numbers with regrouping, or two 4-digit numbers; subtraction—up to 3-digit numbers with regrouping; multiplication—up to 3-digit by 1-digit numbers with regrouping; division—up to 3-digit by 1-digit numbers with and without remainder.	Unit 1 Unit 3 Unit 4 Unit 5 Pretest Posttest	6 1, 2, 3, 19, 20 6, 8, 10, 18, 19, 21, 22 13, 15, 17, 18, 22, 23 3, 7, 8, 9 3, 7, 8, 9
6.4.17 (B/C)	Round whole numbers through the millions to the nearest ten, hundred, thousand, ten thousand, or hundred thousand in contextual problems.	Unit 1	9
6.4.18 (A)	Establish benchmarks (well known numbers used as meaningful points of comparison) for whole numbers, decimals, and fractions (e.g., $\frac{1}{2} = 0.5$, $0.25 = \frac{1}{4}$).	Unit 7	12
6.4.19 (B/C)	Use estimation to verify the reasonableness of calculated results.	Unit 1 Unit 4	10 1
6.4.20 (A)	Terminology: Know that in $q = x \div y$, q is the quotient; in $x + y = s$, s is the sum; in $x - y = d$, d is the difference; in $(x)(y) = p$, p is the product.	Unit 1	13
6.4.21 (B/C)	Use the inverse relationship of multiplication and division to compute and check results. Use these relationships to solve problems (e.g., $5 \times 3 = 15$ and $15 \div 3 = $ ___).	Unit 3 Unit 5 Pretest Posttest	4, 5, 21 16, 19, 20, 21 13 13
6.4.22 (A)	Recognize and write numerals from dictation up to 9,999.	Unit 1	11
6.4.23 (A)	Order whole numbers up to 999.	Unit 1	14
6.4.24 (A)	Recognize equivalent representations of whole numbers and generate them by composing and decomposing numbers through the use of expanded notation to represent numbers (e.g., $3,206 = 3,000 + 200 + 6$).	Unit 1 Pretest Posttest	15, 16 6 6
6.4.25 (A)	Represent and order whole numbers between 0 and 9,999 using symbols (>, <, or =) and words (greater than, less than, or equal to).	Unit 1	17
6.4.26 (B/C)	Use whole number multiplication and division (know the multiplication tables through 12×12).	Unit 3 Unit 4	6, 7, 8 11, 12, 13, 14, 15
6.4.27 (A)	Divide regions or sets to represent a fraction; name and write the fractions represented by a given model (area/region, length/measurement, and set). (Fractions will include halves, thirds, fourths, and tenths.)	Unit 7 Pretest Posttest	13, 14, 19 15 15
6.4.28 (B/C)	Solve addition and subtraction number sentences and word problems with numbers up to 3 digits.	Unit 1	18, 19, 26

Practice for the ISAT Test   **Standards Correlation**

MATHEMATICS ASSESSMENT FRAMEWORK

Objective Number	Objective	Test Item Location	Test Item Number(s)
6.4.29 (B/C)	Add and subtract with decimals expressed as tenths, using pictorial representations and monetary labels.	Unit 8	14

Goal 7: Measurement

Objective Number	Objective	Test Item Location	Test Item Number(s)
7.4.01 (A/C) (OR)	Measure lengths to the nearest $\frac{1}{2}$ inch and $\frac{1}{2}$ cm with a ruler.	Unit 8	15
7.4.02 (A/C)	Compute elapsed time in compound units (e.g., 1 hour and 30 minutes).	Unit 2 Pretest Posttest	1, 2, 3 21 21
7.4.03 (A/C) (OR)	Convert both ways within systems without conversion charts (e.g., yards to feet, feet to inches, meters to centimeters, and hours to minutes).	Unit 8	16, 17
7.4.04 (A/C)	Solve problems that require a knowledge of the following units: inches—down to $\frac{1}{2}$, $\frac{1}{4}$ feet and ?; feet, yards, miles, millimeters, centimeters, meters, and kilometers; weight/mass—ounces, pounds, tons, grams, and kilograms; liquid volume—cups, pints, quarts, gallons, milliliters, and liters; area—square units; temperature (Celsius and Fahrenheit units).	Unit 6 Unit 8 Pretest Posttest	3 18, 25 25, 44 25, 44
7.4.05 (A/C) (OR)	Calculate the area and perimeter of a rectangle, triangle, or irregular shape using diagrams, models, and grids or by measuring. Use the appropriate units in the response [e.g., square centimeter (cm^2), square meter (m^2), square inch (in^2), or square yard (yd^2)].	Unit 9 Pretest Posttest	1, 2, 3, 4, 5, 6, 15, 16, 17 23, 30 23, 30
7.4.06 (B)	Choose the appropriate units (metric and U.S.) to estimate the length, liquid volume, and weight/mass of given objects.	Unit 8 Pretest Posttest	19 29 29
7.4.07 (B)	Estimate the relative magnitudes of standard units (e.g., mm, cm, m).	Unit 8	20
7.4.08 (B)	Estimate standard measurements of length, weight, and capacity.	Unit 8 Pretest Posttest	21 26 26
7.4.09 (A/C)	Perform simple unit conversions within a system of measurement (e.g., feet to inches, yards to feet).	Unit 8 Pretest Posttest	22, 23 27 27
7.4.10 (A/C)	Read temperature to the nearest degree from a Celsius thermometer and a Fahrenheit thermometer (does not require converting between °F and °C).	Unit 6 Pretest Posttest	1, 2 22 22

Goal 8: Algebra

Objective Number	Objective	Test Item Location	Test Item Number(s)
8.4.01 (A)	Know and extend a linear pattern by a well-defined rule or find a rule that fits the pattern (e.g., show a table that pairs number of horses with the number of legs calculated by counting by 4s or by multiplying the number of horses by 4).	Unit 1 Unit 3 Pretest Posttest	22, 23, 24 9, 10, 17 31, 34 31, 34
8.4.02 (A)	Identify or represent situations with well-defined patterns, such as rate of change, using words, tables, and graphs (e.g., represent in a bar graph the growth over five weeks of a plant that grows 1 inch per week).	Unit 2 Pretest Posttest	10 33 33

Standards Correlation   **Practice for the ISAT Test**

MATHEMATICS ASSESSMENT FRAMEWORK

Objective Number	Objective	Test Item Location	Test Item Number(s)
8.4.03 (C/D)	Determine values of variables in simple equations (e.g., $41 - y = 37$, $5 = m + 3$, and $c - 1 = 3$).	Unit 1 Unit 3 Unit 4 Pretest Posttest	20 11, 12 7, 9 28 28
8.4.04 (A)	Solve simple problems concerning the functional relationship between two quantities (e.g., calculate the total cost of several items given the unit cost).	Unit 4	16, 17, 20
8.4.05 (C/D)	Use parentheses to indicate which operation to perform first when writing expressions containing more than two terms and different operations.	Unit 3 Pretest Posttest	13, 14, 22 35 35
8.4.06 (C/D)	Interpret and evaluate mathematical expressions that use parentheses.	Unit 3 Pretest Posttest	15, 16 36 36
8.4.07 (A)	Demonstrate an understanding of equality by recognizing that "=" links equivalent quantities (e.g., $4 \times 3 = 2 \times 6$).	Unit 1 Unit 3	21 18
Goal 9: Geometry			
9.4.01 (A)	Identify, describe, and classify common three-dimensional geometric objects [e.g., cube and rectangular solids (prisms), sphere, pyramid, cone, and cylinder].	Unit 9	8, 13
9.4.02 (A)	Identify regular and irregular polygons.	Unit 6 Pretest Posttest	4, 5, 9, 13, 16 37 37
9.4.03 (A)	Identify the parts of a circle (radius, diameter, and circumference).	Unit 6	7, 8
9.4.04 (B)	Differentiate between polygons and non-polygons.	Unit 6	6
9.4.05 (B)	Identify common solid objects that are the components needed to make a more complex solid object.	Unit 9	9, 10, 11
9.4.06 (B)	Identify prisms (including cubes) and pyramids in terms of the number and shape of faces, edges, and vertices.	Unit 9	7, 12, 14
9.4.07 (A)	Identify paths and movements using coordinate systems.	Unit 6 Pretest Posttest	10 39 39
9.4.08 (A)	Graph points and identify coordinates of points on the Cartesian coordinate plane (quadrant I only).	Unit 6 Pretest Posttest	11 40 40
9.4.09 (A)	Identify, describe, and classify polygons (including triangles, squares, rectangles, pentagons, hexagons, octagons).	Unit 6 Pretest Posttest	14, 15 38 38
9.4.10 (A)	Determine the distance between two points on the number line in whole numbers.	Unit 6	12

Practice for the ISAT Test   **Standards Correlation**

MATHEMATICS ASSESSMENT FRAMEWORK

Objective Number	Objective	Test Item Location	Test Item Number(s)
Goal 10: Data Analysis, Statistics, and Probability			
10.4.01 (A/B)	Use information from a pictograph, bar graph, line graph, or a chart/table, with no more than two variables.	Unit 2 Pretest Posttest	5, 8, 9, 11, 19, 21 41 41
10.4.02 (A/B)	Match a data set to a graph and vice versa.	Unit 2	4, 6, 7
10.4.03 (A/B)	Identify different representations of the same data.	Unit 2	14
10.4.04 (A/B)	Determine minimum value, maximum value, range, mode, and median for a data set with an odd number of data points.	Unit 2 Pretest Posttest	12, 13, 15 42 42
10.4.05 (C)	Classify events as certain, more likely, or less likely by experiments using objects such as counters, number cubes, spinners, or coins, where visual cues are unambiguous.	Unit 7 Pretest Posttest	15, 16, 17 32 32
10.4.06 (A/B)	Use information from a pictograph, bar graph, or a chart/table to answer questions about a situation (which assumes only one variable).	Unit 2 Pretest Posttest	16, 17, 18, 20 24 24

Standards Correlation   **Practice for the ISAT Test**

Name _____   Date _____

Test Answer Sheet   Test Title _____

1. Ⓐ Ⓑ Ⓒ Ⓓ Ⓔ
2. Ⓐ Ⓑ Ⓒ Ⓓ Ⓔ
3. Ⓐ Ⓑ Ⓒ Ⓓ Ⓔ
4. Ⓐ Ⓑ Ⓒ Ⓓ Ⓔ
5. Ⓐ Ⓑ Ⓒ Ⓓ Ⓔ

6. Ⓐ Ⓑ Ⓒ Ⓓ Ⓔ
7. Ⓐ Ⓑ Ⓒ Ⓓ Ⓔ
8. Ⓐ Ⓑ Ⓒ Ⓓ Ⓔ
9. Ⓐ Ⓑ Ⓒ Ⓓ Ⓔ
10. Ⓐ Ⓑ Ⓒ Ⓓ Ⓔ

11. Ⓐ Ⓑ Ⓒ Ⓓ Ⓔ
12. Ⓐ Ⓑ Ⓒ Ⓓ Ⓔ
13. Ⓐ Ⓑ Ⓒ Ⓓ Ⓔ
14. Ⓐ Ⓑ Ⓒ Ⓓ Ⓔ
15. Ⓐ Ⓑ Ⓒ Ⓓ Ⓔ

16. Ⓐ Ⓑ Ⓒ Ⓓ Ⓔ
17. Ⓐ Ⓑ Ⓒ Ⓓ Ⓔ
18. Ⓐ Ⓑ Ⓒ Ⓓ Ⓔ
19. Ⓐ Ⓑ Ⓒ Ⓓ Ⓔ
20. Ⓐ Ⓑ Ⓒ Ⓓ Ⓔ

21. Ⓐ Ⓑ Ⓒ Ⓓ Ⓔ
22. Ⓐ Ⓑ Ⓒ Ⓓ Ⓔ
23. Ⓐ Ⓑ Ⓒ Ⓓ Ⓔ
24. Ⓐ Ⓑ Ⓒ Ⓓ Ⓔ
25. Ⓐ Ⓑ Ⓒ Ⓓ Ⓔ

26. Ⓐ Ⓑ Ⓒ Ⓓ Ⓔ
27. Ⓐ Ⓑ Ⓒ Ⓓ Ⓔ
28. Ⓐ Ⓑ Ⓒ Ⓓ Ⓔ
29. Ⓐ Ⓑ Ⓒ Ⓓ Ⓔ
30. Ⓐ Ⓑ Ⓒ Ⓓ Ⓔ

31. Ⓐ Ⓑ Ⓒ Ⓓ Ⓔ
32. Ⓐ Ⓑ Ⓒ Ⓓ Ⓔ
33. Ⓐ Ⓑ Ⓒ Ⓓ Ⓔ
34. Ⓐ Ⓑ Ⓒ Ⓓ Ⓔ
35. Ⓐ Ⓑ Ⓒ Ⓓ Ⓔ

36. Ⓐ Ⓑ Ⓒ Ⓓ Ⓔ
37. Ⓐ Ⓑ Ⓒ Ⓓ Ⓔ
38. Ⓐ Ⓑ Ⓒ Ⓓ Ⓔ
39. Ⓐ Ⓑ Ⓒ Ⓓ Ⓔ
40. Ⓐ Ⓑ Ⓒ Ⓓ Ⓔ

41. Ⓐ Ⓑ Ⓒ Ⓓ Ⓔ
42. Ⓐ Ⓑ Ⓒ Ⓓ Ⓔ
43. Ⓐ Ⓑ Ⓒ Ⓓ Ⓔ
44. Ⓐ Ⓑ Ⓒ Ⓓ Ⓔ
45. Ⓐ Ⓑ Ⓒ Ⓓ Ⓔ

46. Ⓐ Ⓑ Ⓒ Ⓓ Ⓔ
47. Ⓐ Ⓑ Ⓒ Ⓓ Ⓔ
48. Ⓐ Ⓑ Ⓒ Ⓓ Ⓔ
49. Ⓐ Ⓑ Ⓒ Ⓓ Ⓔ
50. Ⓐ Ⓑ Ⓒ Ⓓ Ⓔ

Practice for the ISAT Test   **Answer Sheet**